T0012947

STATISTICS 101

STATISTICS 101

FROM DATA ANALYSIS AND PREDICTIVE MODELING TO MEASURING DISTRIBUTION AND DETERMINING PROBABILITY, YOUR ESSENTIAL GUIDE TO STATISTICS

DAVID BORMAN

Adams Media
New York London Toronto Sydney New Delhi

Adams Media
An Imprint of Simon & Schuster, Inc.
100 Technology Center Drive
Stoughton, MA 02072

First Adams Media hardcover edition December 2018

ADAMS MEDIA and colophon are trademarks of Simon & Schuster.

For information about special discounts for bulk purchases, please contact Simon & Schuster Special Sales at 1-866-506-1949 or business@simonandschuster.com.

The Simon & Schuster Speakers Bureau can bring authors to your live event. For more information or to book an event contact the Simon & Schuster Speakers Bureau at 1-866-248-3049 or visit our website at www.simonspeakers.com.

Graphics by Chris Monahan

Manufactured in the United States of America

5 2022

Library of Congress Cataloging-in-Publication Data
Borman, David, author.
Statistics 101 / David Borman.
Avon, Massachusetts: Adams Media, 2018.
Series: Adams 101.
Includes index.
LCCN 2018037190 (print) | LCCN 2018039029 (ebook) | ISBN 9781507208175 (hc) | ISBN 9781507208182 (ebook)
Subjects: LCSH: Statistics. | Mathematical statistics.
Classification: LCC QA276.12 (ebook) | LCC QA276.12 .B6678 2018 (print) | DDC 519.5--dc23
LC record available at https://urldefense.proofpoint.com/v2/url?u=https-3A__lccn.loc.gov_2018037190&d=DwIFAg&c=jGUuvAdBXp_VqQ6t0yah2g&r=eLFfdQgpHVW0iSAzG8F-WtSjrFvCD9jGMJBHtzyExXhmHvwB7sjMCnFuKz95Uyqa&m=i_4-sqUtYR1dsHotRZbeuBf23AV9QMee2LEcIqKeRXA&s=A_1aSLk-vo4P_O-X8oNWS-w-ue_0pyuj7s-lkfQSXrY&e=

ISBN 978-1-5072-0817-5
ISBN 978-1-5072-0818-2 (ebook)

CONTENTS

INTRODUCTION

Data analysis, predictive modeling, and *data science*: these terms are splashed across the news, in books, and online. They're applications of a field of mathematics called statistics. Although the names of the applications are varied and diverse, the study and knowledge of statistics is at the core of these disciplines.

Statistics is the measuring of data and interpreting that data to prove or disprove a point. That's it! Statisticians collect and work with large amounts of numbers and measurements. Then they calculate how the numbers relate to each other or how they affect each other.

If you know the basics of statistics—what the numbers tell you—and learn how to collect or obtain accurate, reliable, and good numbers-related facts, and if you further learn to use statistics to arrive at conclusions, then you are well on your way to developing a meaningful statistics-based analysis. This skill can be of great help to you at work, at school, or in your daily life.

This book is for you if you're new to statistics. I'll help you build your knowledge gradually. First, we'll look at core concepts, with each section covering a basic unit of knowledge of statistics. Each section is designed to stand alone, giving you the option of reading the book cover to cover for a comprehensive study. Once you've grasped the basics (or if you already know them), you can jump to a section you're interested in and read only the material you need to learn.

The book can be used if you're studying statistics in a high school or college course. It can also help you brush up if you've take a statistics class in the past, are now being exposed to statistics at work, and would like to know more without taking a full-length course on the subject.

Finally, there are some sections written for high school and college students on how to use statistics in school projects. These units are there to help you understand, design, and test what are called the *quantitative research* sections of assignments, sections that require a statistical analysis.

Read through the book in the order of the sections provided or jump to the section you'd like to learn: either way, you're well on your way to gaining a full grasp of what statistics is and how it's used in business, scientific, and academic studies!

THE BASICS OF STATISTICS

A Tool of Measuring

The science of statistics is used to analyze large groups of numbers. It can be used with spreadsheet software to build simple programs—these programs are called predictive models, and they can help do just that: use statistics to predict the most probable future outcomes of a set of circumstances. While predictive models, data analysis, and data science are different, they all use the same related statistical tools. With them, you'll be able to get data and then provide a testing method to answer a variety of questions. Fundamentally, most statistical studies and models want to know how the numbers that make up the data relate to each other. What's the average? How does the shape of the data look on a graph? And, possibly the most important in any study or research assignment: what does the data tell us?

STUDIES USE STATISTICS

The desktop computer became a reality in the 1980s. People were encouraged to get a desktop computer and were told that it would release users from mundane tasks, giving them more personal time. Well, that didn't happen. What did happen is that governments, companies, and individuals found that there was a great deal of data available to help these entities make decisions. Just like the classic question "What came first—the chicken or the egg?", you can argue the issue as "What came first, the data or the question?" In other words, did the data show that there is a question to be answered or are there questions that require data collection?

Unlike the chicken/egg conundrum, the answer to the data collection/study question is that they are both right. With all the data that is available, people can sit at their desks, analyze the data, and pose questions implied by it or answer questions posed by someone else. Before the age of the desktop computer, that work was being done, but only by those who had access to larger computers and who could pay the bill for the very expensive computer time. (In those days, computers sometimes took up rooms or even, in some cases, entire buildings.)

Today, because powerful computers are accessible and easier to use, statistics have worked their way into the fabric of our daily lives. In today's political campaigns, statistics are an essential tool used by candidates. How many people voted for your political party in the last election? How many of them are men, and how many are women? By what percentage did women support your party? What percentage of the men? Pollsters—a central part of campaign personnel—are experts in statistical analysis. You've no doubt seen a news report in which candidate X leads candidate Y by 5 percentage points (with a margin of error of ± 3 percent).

What's the Question?

With any good study, the question at the study's center must be clearly stated. From there you'll know where to look for good information, and you'll also know what kind of information will help you answer the question. After you've collected the data, you can start using statistics to help find the answer.

Here is a typical question that data analysts use statistics to answer: how many people visiting your company's social media page are in turn visiting your company's website? Once they're there, what advertising blurbs led to the biggest sales?

There is an underlying question here: what's the most effective advertising combination for the biggest sales? The question is embedded within the study.

Finding the right information and data is critical to a good study. Good statistics can be made into great statistics if you can use information and data that are most relevant to the question. In this example, if you successfully use the science of statistics to measure accurately the relation of the websites and advertisements as they relate to sales, you would also be able to build a predictive model that could tell you how the ads would work in the future.

What Is the Value of a Predictive Model?

A predictive model uses a study's results and then builds a tool that gives a good chance of predicting the future with similar data. These models are used in marketing, finance, and medicine, among other fields.

STATISTICS IN SCIENTIFIC TERMS

In more scientific terms, statistics measures the frequency, distribution, randomness, and cause/effect relationship of data points in studies. Statistics is used to determine measures of center, spread, and relative frequency and to create models used for predicting outcomes in finance, marketing, manufacturing, and medicine. It is even used in sports. Michael Lewis wrote a book titled *Moneyball: The Art of Winning an Unfair Game*, which later became the movie

Moneyball, starring Brad Pitt. It describes how the general manager of the Oakland Athletics used sabermetrics (a branch of statistics that deals specifically with baseball) to help a small-market baseball team compete with teams that had more money to spend. Using statistics is becoming more prevalent in the sporting world, and not just baseball. Turn to your favorite Internet sports page, and you'll see row after row of statistics about every sport being played around the world.

When you use statistics, you are looking at groups of numbers from surveys and studies and then measuring how the numbers are related to each other. Finally, statistics can be used to develop a predictive model, with specialized tools that can help determine the cause/effect relationship between inputs of data.

DATA IS THE KEY TO STATISTICS

There are a few basic steps to any statistical study, but they all revolve around numbers, measurements, opinion polls, sales figures, medical study outcomes, stock or other financial trading numbers, etc. The sources of the data can vary widely.

Here's a typical example of the use of data in a study: an educator is trying to determine the optimal factors that prepare eleventh graders for the SATs. She measures high school course load, prior high school college prep grades, hours spent in school-sponsored SAT preparation courses, hours spent in SAT self-study, student hours spent in outside school employment activities, if either or both parents attended college, and number of semester hours and levels of math and English each student has had.

The researcher asks the students and their parents to provide information on these points. This information, once it's collected, is called the data. The researcher uses statistics to measure what can be attributed to most helping a student achieve the highest SAT scores. In other words, the researcher is trying to answer this question: "What are the strongest influencers to my students achieving high SAT scores?"

The Uses of Statistics

Here's a list of some fields in which extensive use is made of statistics tools:

- Stock trading
- Marketing
- Internet sales
- Weather prediction
- Professional sports
- Politics
- Medical research
- Government economic reports
- Advanced academic studies (research papers)

This example highlights the importance of having accurate data. If the answers the parents and students give are wrong—perhaps students exaggerate the number of hours they study for the SAT or parents lie about attending college—the conclusions the researcher draws will be wrong. On the other hand, if the data she's working with is right, statistical analysis will give her the answers she needs.

HOW STATISTICS ARE USED

Statistics Terms and Their Function

Before we get into the details of using statistics, you'll need to learn a few specialized words and terms. Getting to know the technical words and what they mean can help you understand statistics. This section will introduce how statistics are used and some of the key words that you'll need to know going forward.

SOME KEY WORDS

When describing the world we use simple and complex words. While simple words are usually easier to understand, they are actually more difficult to understand when used in the context of statistics. Why is this? Because the complex technical terms that are used to describe statistics can also be used as a sort of shorthand to get at big ideas. By using one or two technical terms, researchers can make complex ideas seem simpler.

For example, words such as *data* imply large amounts of numerical results—most often obtained from a survey or other research. Other words, such as *study*, refer to an entire start-to-finish statistics project. *Observations* relates to the data: how it is collected, how questions in a survey are formulated, and so forth.

STATISTICS TERMS TO KNOW	
TERM	**DEFINITION**
Data	Individual bits of numerical observations
Population	The group containing all possible entities of concern
Sample	A part of the population from whom data is collected
Observation	Each separate collection of one bit of data
Study	Collecting data and using statistics to make an inference about it
Inference	An educated, statistically supported "guess" about a group of data
Descriptive statistics	Values that describe (e.g., center, spread, shape) data sets
Inferential statistics	Making educated guesses, testing theories, modeling observations' relationships, and predicting outcomes with data analysis
Descriptive observations	Data that describes qualities rather than amounts (such as hair color, eye color, etc.)
Random variables	Numerical or descriptive observations that happen by chance
Data set	A group of collected or observed data bits (or data points)
Quantitative data	Data that is numerical
Qualitative data	Data that is not numerical

COLLECTING DATA IS THE FIRST STEP

The first step in all statistics is determining a design for collecting the data. In order to crunch the numbers, you'll need good, reliable data. In fact, the collection of the data for any study can be the most critical factor in finding valid results. It is often unfeasible, if not impossible, to have every element in the population give input to the collection. Therefore, the process for collecting data requires that enough data points be collected without any bias. Asking the question "Who is going to win the World Series?" only in Boston holds great potential for a biased response (because in Boston the Red Sox are *always* going to win the World Series). Ignoring gender, political affiliation, economic status, and other demographic considerations can certainly lead to data that is unreliable.

A critical factor in the accumulation of data is the issue of randomness. One of the methods used in the design of how data is collected to help ensure that the data is not biased is some form of random selection. When conducting a telephone survey, for instance, those conducting the survey use random number generators to determine which telephone numbers from a given region they'll call. High school students, in an attempt to determine how the student body feels about an issue, might ask every third student who enters the cafeteria to complete a questionnaire.

Once the collection process has been designed, you can get a statistically valid sample of, say, one hundred or two hundred data points, which will give the same basic information as one hundred thousand data points. This process is called *sampling* a population. *Sample* is the word for the one or two hundred (the small group), and *population* is the word for the one hundred thousand (the entire group).

Sampling is a key tool in polling. When pollsters say that 72 percent of the population approves of an action by a politician, they don't mean that they asked every person in the country what they thought of the action of Senator Smith. Rather, they developed a representative sample of the state population—representative in terms of race, gender, age, income level, and so on. That's the sample they polled, and they extrapolated from there. If 72 percent of their sample approves of the job Senator Smith is doing, and if the sample is typical of the entire state's population, it's a fair assumption that approximately 72 percent of the voting population agrees that Senator Smith is doing a good job. Of course, such a statistic is only approximate—there's room for error, called the *margin of error.* In later sections we'll discuss how big or small this error might be and how to determine it.

STATISTICS DESCRIBE DATA

From the sample set you will be able to use statistics to characterize the data collected. You will be able to describe the smallest, the largest, the middle, and the most common number in the group. You will also be able to describe how close most of the data points are to the middle. Why is this important? Because you might need to know more than just the average. You may need to know how often an observation (or event, or test, or bit of data) happens, and when it does happen, what the chances are of it happening near its average.

This is a classic example of descriptive statistics. It can go a long way in helping you use statistics to see how often something will occur.

HOW DESCRIPTIVE STATISTICS ARE USED

Let's say a TV station is trying to predict the weather during a snowstorm. The staff at the TV station would like to know the average snowfall on the date in question; they'd also like to know the average snowfall during snowstorms that last more than twenty-four hours. By accessing US weather databases, they'll be able see thousands of measurements of snowfall across the nation for the past sixty-plus years. But we're talking thousands and thousands of numbers—beyond the ability of the staff to analyze in their very limited time frame.

Because the grouping of data is too large to investigate, the TV station takes a sample of the data: they pull one snowfall report for every fifty recorded. The result is a sampling of the entire database over the past sixty years, even though the staff has only looked at one out of every fifty reports. No matter; this is a statistically valid sample.

The TV station then uses statistical methods to see (with a high percentage of accuracy) how much snow will fall after this twenty-four-hour snowfall. From this, the TV station can further break down the data and predict how much snow will fall every hour.

KEY POINTS OF STATISTICAL ANALYTICS

Using Statistics to Describe, Interpret, and Model

The object of statistical analytics after the collection and interpretation of the data is to interpret this data. In this respect, the size of the data set doesn't really matter, whether it's a sample drawn from a much larger body of information or if the study itself had only a few observations and, therefore, a smaller data set. Either way, after the data is collected it can then be analyzed. This section will discuss the two types of data analysis: descriptive statistical analysis and inferential statistical analysis.

DESCRIPTIVE ANALYTICS

Descriptive analytics is the measuring, sorting, and study of data and the process of describing it. When you first look at a set of data, you can tell a lot: the largest number, the smallest number, the average number, and so on. You can also tell how close around the average number in the middle the set of data is grouped. In other words, you can tell not only the average, but also what percentage of the numbers in your study are close to the average and how close.

This is important in helping you find out how often something happens. In a medical study, you might need to see not only by how much the new medicine lowers a fever, but you might also need to know how frequently it has that result. With descriptive statistics, you can tell not only the average number of degrees by which a fever was reduced but also the range of temperatures the fever was reduced by, say, in more than half of cases. In this example, you would be using descriptive statistics to find not only an average, but also a frequency.

Descriptive or Inferential?

How do you know if you are talking about descriptive or inferential statistics? If you are *describing the data* with measures of center, spread, or shape, then it's descriptive statistics. If you *draw conclusions from the data* to predict center, spread, or shape, then it is inferential statistics.

INFERENTIAL ANALYTICS

A second way that statistics are used is called inferential statistics. Inferential statistics help you make inferences, or educated guesses, about the information contained in a data set. You draw conclusions, although possibly tentative ones, on how pieces of data relate to one another. This is important in creating models that are used to predict future outcomes.

How is this done? First, you use statistics to measure the quality of the data you've collected. You do this to determine if any inferences or educated guesses you made are accurate.

Let's say you want to know what kinds of car maintenance can lead to the highest increase in a car's fuel efficiency, measured in miles per gallon (mpg). If you've made a guess, you'll want to know how accurate it was. You take measurements (called observations), and then you use statistical tools to determine which of the inputs—say, tire pressure, oil changes, quality of gasoline, and outside temperature—had the most effect on the car's fuel consumption. At the same time, those tools will tell you what factors or inputs you observed didn't have any effect at all on the car's mpg. With statistics, the goal is usually 95 percent accuracy before a researcher can assume that the guess is correct. With inferential analytics, you can tell (ideally with 95 percent accuracy) what kinds of car maintenance (the data inputs in the study) lead to the greatest increase in the car's mpg.

With this information you're now ready for theory testing. This is one of the reasons that statistical studies are done in the first place.

More Vocabulary

The three biggest items that statisticians concern themselves with are the measures of center, spread, and relative frequency. When these items measure a population, they are called *parameters*. If they are measures of a sample, they are called *statistics*.

A car company suspects that tire air pressure and frequent tune-ups lead to the highest level of mpg improvement over time. They also suspect that changing the oil more often affects mpg, but they're unsure by how much. After conducting a study, the car company can apply an analysis of variance (ANOVA) to see that, indeed, keeping tires properly inflated relates directly to higher mpg. In fact, they know this with 95 percent certainty. However, more frequent tune-ups directly result in higher mpg with only 75 percent certainty, and more frequent oil changes are down to a 50 percent certainty. With this new information, the car company can now take the experiment one step further: they can measure how much tire pressure results in the greatest increase in mpg and stop the tune-up and oil change part of the study. This is an excellent example of the inference part of statistics. (We'll discuss ANOVA in more detail later in the book.)

MODELING

The last, and perhaps the most interesting, part of statistics is modeling. Using software such as Microsoft Excel and other (sometimes more complex) software, the car company is able to take the information they've found about tire pressure and its relationship to mpg and build a predictive model.

The goal of the predictive model is to use the past to predict the future. How is this done? In the tire problem, all the tested tire pressures are measured against the resulting increase in the car's mpg. From this, a computer calculates the predicted mpg of a car if the tire pressure is x.

The Need for Speed

The kind of tire pressure/mpg/speed measurements we've been talking about in this section are done at the Daytona 500, Indy 500, etc. Each race team's pit crew inputs such data as track temperature, air temperature, weather, driver weight, and so forth to find the optimal mix of race car adjustments. In effect, they're saying, "There are things we can't change on race day, such as the weather and temperature. What do *we* control on the car to make it as fast as possible?"

MIXING UP THE TEST

Randomness and Random Sample Sets

As we've seen, when available data is impractically large for purposes of analysis, you can test a smaller part of the whole. This is called a random sample set.

This smaller sampling technique works just fine, if you follow certain rules. The main rule is to maintain a random sample of the larger population. Randomness is the key to sampling.

DIFFERENT WAYS TO RANDOMLY TEST

There are several different ways to make sure that your smaller sample set represents a true example of the larger sample set. *Systematic sampling* means that you test one out of every specified number of samples, regardless of what order they are in. For instance, if you were testing coffee drinkers at a café, you could test every hundredth customer who came into the store. This sample set would be random, because, all other things being equal, you would have no way of knowing who was going to walk through the door next.

The second way is *stratified sampling.* To make your sample set even more random, you separate the chosen number into categories—say, men and women. In the case of the coffee shop, you survey every hundredth man and every hundredth woman. You then combine the results, offering an even greater level of diversification and randomness in your sample set.

You can obtain additional levels of randomness by *cluster sampling,* which in this case means sampling coffee customers at randomly chosen locations. Finally, there is *convenience sampling*. Here you'd only address questions to coffee customers when and where you buy your own coffee, with no set pattern and no set times but strictly at your own convenience. (This last method is the least random.)

MISLEADING RANDOM SAMPLE SETS

Random sampling is widely used when the entire population of the study is large. This is typically the case in presidential election polling. However, there are times when such polling can be highly misleading. A classic example of this was in 1936, when Republican candidate Alf Landon and Democratic candidate Franklin Roosevelt were running for the presidency. A popular magazine, *The Literary Digest*, was commissioned to perform the largest-ever presidential election poll. In this poll, the magazine sent out 10 million questionnaires, asking readers whom would they vote for. The magazine received 2.1 million ballots; these showed that Landon would get 57 percent of the votes and would win the election.

Of course, anyone who's read history knows Roosevelt won the election. What went wrong with this sample set? Surely its huge size should have led to accurate testing.

The problem wasn't with the size of the sample but rather the fact that it wasn't random enough. The magazine sent out the poll questions to magazine subscribers, owners of autos, and those who had a telephone in the home. This created an immediate bias: remember, in 1936 the United States was in the middle of a deep economic depression, and those who subscribed to magazines, owned cars, and had phones in the home were typically better off financially than those who didn't. In fact, these more affluent people were almost exactly the profile of voters who would vote Republican in the next election. Because of this inherent bias in the polling methods, the sample itself was biased, and the results were false!

Sometimes, even statistics aren't enough to predict an accurate outcome. In the 2016 US presidential election, polls showed Democrat Hillary Clinton with a comfortable lead for much of the campaign. Even toward the end, when the race between Clinton and Republican Donald Trump narrowed significantly, Clinton's campaign advisers clung to their belief in the polls (and polls in 2016 were certainly much more sophisticated than those in 1936).

Kennedy's Polling

The first US presidential campaign to make use of a private poll was that of John F. Kennedy in his 1960 run against former vice president Richard Nixon. Since then, virtually all serious presidential, senatorial, and congressional campaigns have made extensive use of private polling services.

As we all know, Clinton lost the election, and Trump was elected president. What happened? Pundits are still debating, but it seems clear that two factors influenced the election. First was the revelation by the FBI that it was investigating certain of Clinton's emails, focusing (or rather, refocusing, since this issue had come up earlier) the public's attention on her use of a private email server while she was secretary of state. Second, her campaign made a decision to campaign minimally in what turned out to be battleground states.

Trump's campaign won these states and, subsequently, the electoral college. For all of the sophisticated polling methods used, both campaigns were surprised at the outcome.

KNOWING THE QUALITY OF YOUR DATA

Is Your Information Good?

The starting point of any statistical study is the collection of data. The data can be sourced from government, financial, or medical databases. The information can also come from online surveys, personal interviews, or mailed questionnaires.

SurveyMonkey

One of the best ways to perform electronic interviews is to set up your questionnaire through the survey company SurveyMonkey (www.surveymonkey.com). The company helps you build your surveys and then sends them out to its subscriber list.

After you've gotten your survey ready, sent it out, and received your responses, the next thing to do is see how good your responses are.

THE QUALITY OF YOUR DATA

One of the best ways to determine the quality of your data is to first measure how many responses you've received. This is going to depend on how large your entire population is. Ideally, if you're sending out a questionnaire or taking an online survey, you'd like to get a response from a minimum of 2 to 5 percent of the target population. Of course, the higher the response rate, the better the quality of the data.

Second, look at the range of responses you're getting. Are the responders from diversified backgrounds? This might refer to factors of geography, economic status, political opinions, age, or educational background. A diversity of responders is a contributor to good data.

Remember, you'll most likely never get a 100 percent response rate, so your sample set must be a good cross representation of the entire population. Testing the quality of the sample set means having some qualifying, nonrelated questions in your questionnaire. These questions could be test questions to see if the person taking the survey is somehow biased or otherwise not good for the poll. This may be the case if the person works in an industry that is too closely tied to the study or is politically, socially, or economically biased in a way that would throw off your results. If this were the case, these people's answers wouldn't be included in the study.

You'll need to think of the best ways to ask the blind questions to have them work properly in your surveys. Ask yourself, "What are the factors that could throw off my results?" After working with statistical analysis for a while you'll develop a good idea of what can go wrong with data.

It might be that you're collecting data on whether people prefer to use over-the-counter cold medicine or if they get flu shots during flu season. You start by listing all the things that could skew your results:

- Over-the-counter cold medicine or flu shots
- Younger age or senior citizen
- Climate of responder (snowy regions or sunny regions)

- Income bracket
- Easy access to flu shots at clinics
- Insurance with low deductibles

How would these skew the results? Younger age and senior citizen patients are generally targeted to get flu shots, people in snowy regions might be more favorable to flu shots (since they stand a better chance of getting the flu), higher-income people could more easily afford the flu shot, people with free flu shot access would be more inclined to get flu shots, and people with low insurance deductibles would be more able to afford flu shots. Remember, you are measuring those with no undue advantage versus the others in the study. Responders should be close to being equals in demographics.

After getting the surveys back, you look at the qualifiers to see if the quality of the data is being skewed—that is, has one subgroup responded substantially more than others? If, of course, there isn't a factor that stands out, the quality of your data is much better.

Remember, if you are designing a study, keep track of all your questions, blind questions, response rate percentage, and any blind question qualifiers. If you are writing a report of your study, then this information goes into a section called "The Research and Questionnaire Design."

MODELING RISK, MEASURING SAMPLES, AND PREDICTING

Measuring Good and Bad Outcomes

In order to use statistics properly, you must conduct fact-finding missions to discover the data you will use in your study. Perhaps you send out a questionnaire, then record all the answers that are sent back to you. You can also collect data from databases. Many surveys have been done on various topics, and these databases are usually from governmental, financial, or other reliable third-party sources. If you want to know about people and how they behave, the US Census Bureau is a great source of data (www.census.gov/data.html). If you're looking for financial data, consult *Yahoo! Finance* (https://finance.yahoo.com/) and *BigCharts* (http://bigcharts.marketwatch.com/). All of these are free to use. Other research sites offer vast databases but are subscription based, some even with a one-time use charge. (We'll discuss this more in a later section.)

Once you've collected the numbers and the data that you need for your study, the next thing you need to do is measure how random it is. The tighter around a center the data is grouped, the better its quality, and the more predictable the outcomes will be when you use it to forecast the future.

For example: if you are measuring how many points a National Basketball Association (NBA) team will score after scoring more than 90 points, you might discover that the scores you found in your study are much more meaningful if most of them hover closely around the same average. For instance, you might find that 68 percent of the time NBA teams scored more than 10 additional points after they had scored more than 90 points. If, however, you find that 99 percent of the teams score more than 10 points, you'll reason that the relationship between the two is more predictable. It might not be 100 percent accurate to say they are directly related, but you can say that the data you've collected is tightly centered around the average score. In other words, you have a 99 percent certainty that teams score at least 10 points after they have put up more than 90 points. Ninety-nine percent certainty is excellent and is considered near perfect.

RISK

Risk, on the other hand, is the opposite of certainty. In this case, the risk is that the team will score less than 10 points after the first 90, or possibly 0 points after the first 90. If you've done the same research as you've done for the good outcome, then you can figure out the bad outcome: what's the risk that the team will not score at least 10 more points? In this case, based on the data you've collected, it's negligible. How does this figure?

Look at it this way: there is a 100 percent chance *something* will happen. We don't know what will happen, but something will. There is a 99 percent chance (as our database shows) that the team will score 10 points or more after scoring 90 points. So, if there is a 100 percent chance something will happen, and a 99 percent chance the team will score 10 or more points, then that leaves only a 1 percent chance that the team will score something other than 10 points or more. *This means the risk is 1 percent.*

INDEPENDENCE AND DEPENDENCE

Risk, like success, is a measure of outcomes. Not all measures of risk are measures of negative outcomes. One classic example of a measure of risk is a couple who wants to have two children but wants to have only two girls. What is the measure of outcome that they will have only two girls when having only two children?

The probability that the second child born is a girl knowing that the first child born is a girl is 50 percent. Because the probability that the second born is a girl is not affected by the gender of the first child, the two events "gender of the first child" and "gender of the second child" are independent of each other.

Pick a Card, Any Card

A second example of independence is drawing two cards from a traditional deck of cards. Pick a card from the deck, look at it, put it back in the deck, shuffle the deck, and pick a second card. The probability that the second card drawn is an ace does not depend on the first card drawn. The chances of it being an ace are 4 in 52. However, if the first card is not returned to the deck, then the chances that the second card is an ace depend on the first card selected. If the first card is an ace, the probability that the second card is an ace is 3 in 53, while if the first card was not an ace, the probability that the second card is an ace is 4 in 53.

The first child will have a 50/50 chance of being a boy or a girl. So, the possibility of the first child being a girl is 50 percent. What then are the chances of having a boy or a girl with the next baby? 50/50. So, the first chance of a girl is 50 percent, the second chance of a girl is 50 percent, and the chance of having two girls is 50 percent of 50 percent, or 25 percent. Therefore, there is a 25 percent chance of having two girls when having only two children. A third girl when having only three children would have even less risk of happening: 50 percent for the first child, 50 percent for the second child, and 50 percent for the third child. So, 50 percent of 50 percent of 50 percent, or a 12.5 percent chance (50 divided by 2, divided by 2 again), is the probability of having three girls. Each subsequent birth splits the overall chances by a factor of 50 percent. In this way, you are measuring the risk of outcome of three births/three girls. This is also known as the probability of the three births/three girls outcome. As you see, a measure of risk can also be a measure of probability of an outcome: *it's a measure of the chances of that event happening.*

FREQUENCY DISTRIBUTIONS

How Frequently Observations Occur

By its nature, statistics is concerned with large amounts of data. Sometimes this data is collected with in-person observations, sometimes the data comes from professional or commercial databases, and other times the data comes from government databases. In this section we will talk about the organization of this data into groups, called *frequency distributions*.

Data that is collected needs to be organized in such a way that it can then be used in the actual statistical tests. Large bodies of information you've collected need to be organized into groups so that the tests can be made and, more importantly, so you'll be able to understand the results of those tests you run. Data that is collected is meaningless unless it has been organized. This data from the data collection part of your study can be broken down into two groups:

1. Categorical distributions
2. Numerical blocks of information, or frequency distributions

CATEGORICAL DISTRIBUTIONS

Any study starts with the collection of data. This data comes from observations, and these observations tell a story about what you are studying. Sometimes the story is told with words and sometimes with numbers. Let's say you're analyzing the hair color of college freshmen and whether it affects their grades. (This seems like an odd correlation, but bear with us.) In this case, you (or your data collector) could ask college freshmen what their grade point average is and, while they are answering, make a notation of their hair color. For purposes of the study, you break down hair color as either red, brown, black, or blonde. These four colors of hair are different, yet they are not quantitative—that is, none of them are numerical.

Qualitative Data

It is a universal truth that three is larger than two. Numbers are structured to be compared. However, is red hair "greater" than blonde hair? That's a question about personal preference, not a question that can be universally agreed upon. Other qualitative characteristics along these lines include marital status, eye color, and level of education (indicated by the highest degree earned).

When could you quantify hair color? The answer is simple: in either a marketing campaign or a big sales advertising rollout for a hair-coloring product, you might need to know if some people thought one color of hair was better or more appealing than the other color. Why? Well, if you know the preferred hair color, you can feature it in your TV ads more dominantly. If blondes have more fun, you'll want to concentrate your advertising on people who want their hair to be blonde.

You can then refine this. Perhaps, as a result of the study, you learn that ash blonde is the least popular in your target market and that platinum blonde is the most popular. Therefore, models with platinum blonde hair color get more time in the filming of TV spots that are used to advertise the hair coloring product.

In this case, could you use a number to quantify hair colors? In a survey, you probably could. For instance, you could phrase a question that said, "Rate the following hair colors from 1 to 5: Raven Black (1), Chestnut (2), Mousy (3), Ash Blonde (4), Platinum Blonde (5)." Keep in mind, the assignment of numbers could just as easily have been reversed: Platinum Blonde (1), Ash Blonde (2), Mousy (3), Chestnut (4), Raven Black (5). People will respond to the survey. Those reading the results may select just the top preference, or they may rank the preferences from top to bottom. Are the numbers associated with each color important? Not really. The responders could have written their responses, but data entry with numbers is more convenient. (Consider how much time responders want to spend writing down their responses.)

A frequency chart may be created to examine the results. You could make a bar chart or pie graph to display the outcome. The data being collected is hair color. The summary gives the frequency of responses for this qualitative characteristic.

PREFERRED HAIR COLOR

HAIR COLOR	FREQUENCY
Platinum Blonde	34
Chestnut	56
Ash Blonde	43
Raven Black	12
Mousy	6

Quantitative versus Qualitative

Drill into your mind this distinction; it's important. *Quantitative* refers to numbers-based data. *Qualitative* refers to words-based data. If you remember this, you'll have mastered an important point in statistics theory.

QUANTITATIVE FREQUENCIES

Quantitative frequencies are tables of data that are numbers based. For instance, suppose you're concerned with the problem of wear and tear on a section of highway. You're studying how many cars with four tires pass through a tollbooth each hour of the day, measured over a seven-day period. You want to compare this with how many semitrucks with eighteen tires pass through the tollbooth. Your hypothesis could be as follows:

- More semitrucks pass through the tollbooths in the late evening hours, and more cars pass during rush hour times.
- These numbers give you a number of tires on the road per hour, because that number of tires on the road is tied to the damage done to the surface of the highway.
- There is a sequence to the hours also—late in the afternoon is different than early in the morning. (This could be used to restrict which lanes trucks may use so that the congestion of rush hour traffic can be managed more efficiently.)

Do Red Cars Get Stopped More Often?

There is a persistent belief that red cars are stopped more frequently by the police for speeding. The police deny this is the case, and in 1990 a reporter for the *St. Petersburg Times* in Florida conducted his own statistical study. He found that not only are red cars not stopped out of proportion to their presence in the overall car population, but gray cars seem to be the ones that get stopped more frequently than they should. Statistics busts another myth!

This is all quantitative data; it can be measured in numbers. If the data is reported in a frequency table, the data is numerical.

Keep in mind that if you were, for whatever reason, to observe the colors of the cars passing through the tollbooth, this would be qualitative information.

FINDING THE RELATIVE FREQUENCY DISTRIBUTION

In either type of data, qualitative or quantitative, the results are similar. You will be left with a list of the frequency of each thing happening. While red/blonde/black/brown hair can't be quantified (in the sense of saying that one color is better than another), the number of times it shows up in your total study can be measured. The same is true with the trucks and cars. Finding the relative frequency distribution is simple: take the number of observations of that item and divide by the total number of observations.

FREQUENCY DISTRIBUTIONS OF VEHICLES		
VEHICLE	**FREQUENCY**	**FREQUENCY DISTRIBUTION**
Ford 350	12	12/183 = 0.07
Toyota Camry	70	70/183 = 0.38
Honda Accord	65	65/183 = 0.36
Chevrolet Colorado	15	15/183 = 0.08
Toyota Tundra	21	21/183 = 0.11
Total	183	= 183/183 = 1.00

This table shows the type of vehicle and the total number of times each vehicle was observed. The last column shows the frequency for each vehicle divided by the total number of observations, which gives the relative frequency distribution of each vehicle. Notice that the relative frequency distributions add up to 1. This is the same calculation that would be used to find the frequency distribution of quantitative, or numbers-based, observations.

DISCRETE VERSUS CONTINUOUS DATA

You are a math teacher (don't get nervous, there are a lot of us), and you've just given an exam to the two sections of Algebra I that you teach. You want to see how your students did, so you tally the scores to see how many students scored in the 90s, the 80s, the 70s, etc. You can create a chart with intervals 90–100 (we'll include those who scored 100 because they earned an A on the exam), 80–89, 70–79, 60–69, 50–59, and so on to ensure there is an interval for every exam graded. Since most grades are integer values (if half points are awarded on the exam, the grades can be rounded to the nearest integer), we need not concern ourselves with decimal values. These integer values are discrete data. We can then look at the results to assess the level of difficulty of the exam. We'll be able to determine where the center of the data is located as well as the spread of the data. Should we choose, we can make a graph of the outcomes.

A sample frequency chart might look like this:

FREQUENCY DISTRIBUTIONS OF TEST SCORES	
SCORE	**FREQUENCY**
90–100	17
80–89	19
70–79	21
60–69	10
50–59	2

In a different scenario, you are the quality control manager for a manufacturer of a company that makes ball bearings. You order the staff to randomly select 250 ball bearings from the latest manufacturing process and measure the diameters of the ball bearings with accuracy to the nearest tenth of a centimeter. If the ball bearings are designed to have a diameter of 2.0 cm, your intervals might be 2.0–2.1, 2.1–2.2, 2.2–2.3, etc., for the measurements that exceed the specified value of 2.0 cm. "Wait!" you say. "What do I do if the measurement is 2.1 cm?" In situations like this, the first interval really is $2.0 \leq$ diameter < 2.1, $2.1 \leq$ diameter < 2.2. In this way, if the diameter measures 2.11 cm, which is larger than 2.1 cm, we have a clear definition for which interval the item should be included in.

FREQUENCY DISTRIBUTIONS OF DIAMETERS OF THE BALL BEARINGS	
DIAMETER	**FREQUENCY**
1.7–1.8	2
1.8–1.9	1
1.9–2.0	10
2.0–2.1	210
2.1–2.2	27

DOT PLOTS, BAR CHARTS, HISTOGRAMS, FREQUENCY POLYGONS

The Data Tells a Story

Sometimes it's best to look at the data in a statistical problem with more than just numbers: sometimes you need to look at it with shapes, charts, and graphs. This section will cover introductory graph types used in statistics—dot plots, bar charts, histograms, and frequency polygons.

DOT PLOTS

A dot plot is a type of chart that has one row of information across the bottom and another row of information along the side of the chart. This information shows the different observations along the bottom, with the frequency or number of times that event happened indicated by dots stacked along the chart.

Observations

In statistics, when an event happens it is called an *observation*.

Each observation is measured in units across bottom of the dot plot, and the number of times each separate type of observation happened is shown in dots stacked up from the bottom. One dot usually signifies each time that type of event was observed or happened.

SALES CALLS

Let's consider a study that observes how many times a sales team makes cold calls to potential clients and how many of these calling patterns result in an appointment for a presentation in the client's office.

NUMBER OF CALLS	NUMBER OF SALES LEADS
2	1
4	3
6	5
8	6
10	1
12	0
14	1
16	1
18	0
20	1

The sales team has made cold calls from 4 to 20 times to each potential customer in the hopes of gaining an agreement for an in-office presentation. With 2 calls, an average of 1 appointment is made. If the sales rep makes 4 calls, the result is 3 appointments. With 8 calls, 6 appointments are made. However, at this point something strange starts to happen. Fourteen, 16, and 20 calls resulted in only 1 appointment each. From these observations, and by plotting the data on a dot plot chart, what would a sales manager conclude about the effectiveness of cold-calling potential customers?

A dot plot of this sales call chart would show that most appointments are made when the rep makes between 4 and 8 calls to the potential client. It also would show that only 2 calls are not that effective at gaining appointments, nor are additional calls beyond 8 calls. Based on this, the sales manager tells his sales team to cold-call their potential clients a minimum of 4 times, but if the appointment isn't made by the eighth call, to abandon the lead, since there is a statistically significant chance that the prospect will never set up an appointment and therefore will never become an actual client.

Dot Plots As Foundation

Dot plots are the most basic of the statistics graphs—from this graph you can construct a bar chart, a histogram, and a frequency polygon. This can easily be done with chart and statistics software by toggling among the four chart types.

BAR CHARTS

A bar chart uses the same concept as the point chart, except the data is presented in the chart in solid, vertical bars. In the case of the cold calling sales chart, the chart shows only the totals of the data. A bar chart in the case of the cold calling has the exact same shape as a dot plot, but instead of more dots being used to show more appointments made, a bar is drawn taller for each appointment made.

A bar chart is a good way to display the data of a study that involves the comparative changes in things that are numbers based (i.e., when the observations are not qualitative, such as color, shape, taste, or other similar data). A bar chart works if you are measuring the number of people who wear different shoe sizes but not the number of shoe sizes that are different colors. Shoe sizes are qualitative (that is, numbers based), while the different colors of the shoes are not numbers based but are qualitative in nature.

Bar Graphs versus Histograms

Bar graphs are used to display the frequency of a distinct entity, and thus each bar is separate from the others (as in a chart showing shoe sizes—there are no sizes between an 8 and an 8½), while a histogram shows an interval in a continuum of data (such as test scores), and thus the bars are connected.

HISTOGRAMS

A histogram is a bar chart in which each bar touches another to form a solid mass of bars. The shape provided by the histogram gives a visual representation of the shape of all the data that was collected in the study. Knowing the shape of the data can help show where most of the data is centered. Histograms are great ways to see the shape of the data in a simplified format, without the need for complex charting methods. The histogram indicates where the center of the data can be found, indicates whether the distribution is balanced about a center or skewed to one side or the other, and gives a sense of the spread of the data.

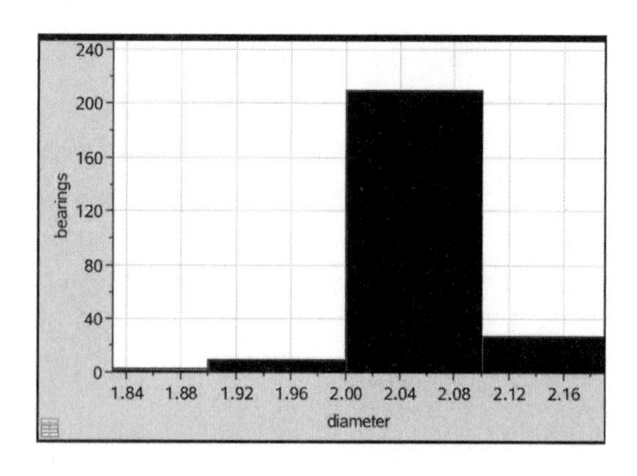

FREQUENCY POLYGON

After you've made a bar chart and then made a histogram, draw a dot at the top of each bar center and then connect these dots with a line. The shape of the line with its up-and-down peaks and valleys is called a frequency polygon. This tool can be used to provide an estimate of the shape of the histogram. The line shows the peaks and the valleys of the histogram, which in turn shows the shape of the chart. These lines provide a level of detail that occupies a space between the histogram—the simplest—to the bell curve, which is more complex. The bell curve also helps show where most of the data is centered. Looking at where the data is centered can help give a visual estimate of the data's average.

Scale

Two crews in a manufacturing process give presentations to the plant manager about the outcome of a project. The frequency polygon for one group shows variation in their day-to-day outcomes, evidenced by the polygon rising and falling from one day to the next, while the polygon for the second team is almost linear. An immediate reaction might be that the second team's performance was fairly consistent. However, the plant manager notices that the vertical scale for the second team is 10 times greater than that of the first team. Consequently, a dip of 5 units is more evident on the graph of the first team than the second team. Pay attention to scale when reading graphs. It is easy to be fooled.

MORE WAYS TO SEE NUMBERS-BASED DATA

Stem and Leaf Plots, Time Series Graphs, Pie Graphs

The importance of choosing the appropriate graphical representation for data is that, like a still photograph, it can give the reader a great deal of information in a short time. The graphic may convey information taken over time, may compare one element to another, or may show the shape of the distribution without losing the data values themselves. This section will cover these three statistical charts and graphs: stem and leaf plots, time series graphs, and pie graphs.

STEM AND LEAF PLOTS

A stem and leaf plot shows data broken down into sections of stems and leaves. By using a stem and leaf plot, you can illustrate the center and spread of the data without losing the data itself. The stem can be either the hundreds or the tens of the numbers, and the leaves can be either the tens or the ones digits of the numbers. In other words, if the number was 113, the stem can be either 1 (if you're using hundreds as the stem) and the leaf 13, represented as (1,13), or the stem can be 11 (if you're using tens as the stem) and the leaf 3, represented as (11,3).

The following table lists the seating capacity, rounded to the nearest hundred people, for the teams in the National Football League (NFL). (Please note that there are only 31 data values because the New York Giants and the New York Jets both play in the same stadium.)

STADIUM	SEATING CAPACITY (IN THOUSANDS)	TEAM
Arrowhead Stadium	76.4	Kansas City Chiefs
AT&T Stadium	80.0	Dallas Cowboys
Bank of America Stadium	75.4	Carolina Panthers
CenturyLink Field	68.0	Seattle Seahawks
FedEx Field	82.0	Washington Redskins
FirstEnergy Stadium	67.9	Cleveland Browns
Ford Field	65.0	Detroit Lions
Gillette Stadium	66.8	New England Patriots
Hard Rock Stadium	65.3	Miami Dolphins
Heinz Field	68.4	Pittsburgh Steelers
Lambeau Field	81.4	Green Bay Packers
Levi's Stadium	68.5	San Francisco 49ers
Lincoln Financial Field	69.6	Philadelphia Eagles
Los Angeles Memorial Coliseum	77.5	Los Angeles Rams
Lucas Oil Stadium	67.0	Indianapolis Colts
M&T Bank Stadium	71.0	Baltimore Ravens
Mercedes-Benz Stadium	71.0	Atlanta Falcons
Mercedes-Benz Superdome	73.2	New Orleans Saints
MetLife Stadium	82.5	New York Giants, New York Jets
New Era Field	71.6	Buffalo Bills
Nissan Stadium	69.1	Tennessee Titans
NRG Stadium	72.2	Houston Texans
Oakland–Alameda County Coliseum	53.3	Oakland Raiders
Paul Brown Stadium	65.5	Cincinnati Bengals
Raymond James Stadium	65.9	Tampa Bay Buccaneers
Soldier Field	61.5	Chicago Bears
Sports Authority Field at Mile High	76.1	Denver Broncos
StubHub Center	27.0	Los Angeles Chargers

STADIUM	SEATING CAPACITY (IN THOUSANDS)	TEAM
TIAA Bank Field	67.2	Jacksonville Jaguars
US Bank Stadium	66.7	Minnesota Vikings
University of Phoenix Stadium	63.4	Arizona Cardinals

Source: "List of Current National Football League Stadiums." *Wikipedia.* https://en.wikipedia.org/wiki/List_of_current_National_Football_League_stadiums.

Look at a stem and leaf chart for the following data: 78, 79, 112, 115, 153. This is a stem and leaf chart that uses the hundreds and tens as stems and units as leaves:

7|8 9
11|2 5
15|3

This is a stem and leaf chart that uses the hundreds as stems and the tens and units as leaves:

0|78 79
1|12 15 53

Let's look at the stem and leaf chart for the seating capacity of NFL stadiums. We include a legend so that the reader can interpret the data. Legend: 7|64 = 76.4 thousand people. Note the stem, 7, is written to the left of the vertical line while the leaf, 64, is written to the right of the vertical line. Next, examine the data and determine the minimum and maximum stems—in this case, 2 and 9. Write these stems vertically on your paper with the vertical lines (the trunk, if you will) next to them.

2|
3|
4|
5|
6|
7|
8|
9|

Work through the data from top to bottom and fill in the leaves. Notice that you do not have to sort the data from least to most ahead of time.

2| 70
3|
4|
5| 33
6| 80 79 50 84 85 96 70 91 55 59 15 72 67 34
7| 46 54 10 10 32 16 22 61
8| 00 20 14 25
9| 36

Turn your book 90 degrees counterclockwise to see the histogram for the data.

Stem and leaf charts are used when you're looking at numbers that can be rounded to two or three significant digits and you'd like to simplify your study. These charts allow you to have a simple way to look at numbers in a more streamlined fashion, and they are used if the study you are conducting doesn't need a very critical accuracy level or if you have a reasonably sized set of data. For instance, if the data consisted of the salaries of the two thousand players who received paychecks for playing in the NFL in 2017, stem and leaf would not be a wise choice for displaying data.

TIME SERIES GRAPHS

If a group of data is collected over time, you can use a time series graph to display the information. Time series graphs use time across the bottom of the graph and show how many times the event happened (how often an event happens is called the *frequency*) along the vertical side of the chart. The time series can be in seconds, minutes, hours, or longer and is spaced out to match the time scale of the study. That is to say, the scale used on the horizontal axis can agree with the intervals when the measurements are taken. The frequency scale on the side of the time series graph should match the number of times the event happened. The graph then lines up a point on the chart by drawing an imaginary line up from the time across to the number of times the event happened. At that intersection, you plot a dot on the graph. After this is done each time, you connect the dots to draw a graph with the peaks and valleys of the graph. This completes the time series graph.

Time series graphs can be very useful when your study shows the progress of observations as they are spread out over time.

An Example of a Time Series Graph

Imagine that an auto manufacturer wants to measure the temperature of an engine as it warms up during a normal operating cycle and then measure how many times the emissions sensor on the car turns on. These types of studies are very common with auto manufacturers as a way of finding out if the emissions sensors within the car's computers work properly. Car manufacturers know that these sensors must measure the fuel/air mix to work properly, yet the car's computer won't turn on unless the operating temperature warms up.

With this study, the manufacturer wants to know, "How many times does the emissions sensor work when the car is warming up—up to the point where the car is fully warmed up?"

Analytic Software

Software is often designed especially for an industry. In the car industry, a car's onboard computers measure the frequency of on/off cycles during different driving conditions. The data is stored in the memory on the car's computer. This data is then analyzed by an auto tech with analytic software to see if the car is running properly under all speeds and temperatures.

The data is collected with the memory of the onboard computers of the car, downloaded to software, and then displayed in a time series graph. Time of operating (getting the car to warm up) goes across the bottom, and the vertical side data points represent the number of times (in scale) the computer turned on the emission sensor. Based on this data, the manufacturer can adjust the sensors for maximum efficiency.

PIE CHARTS AND PIE GRAPHS

A pie chart/pie graph can be an effective tool if the data that was collected in the study has a cumulative value of 100 percent. Since all the data fits in the circle, all the data adds up to 100 percent.

An Example of a Pie Chart

There are ninety-five students in the school district's fifth grade. A school administrator wants to see how many of these students are in five levels of the school's reading program, with each level progressively more difficult than the next.

NUMBER OF FIFTH GRADERS IN EACH READING LEVEL	
READING LEVEL	NUMBER OF STUDENTS
Level 1	8
Level 2	16
Level 3	32
Level 4	23
Level 5	16
Total Fifth Graders	95

Each level of the school's reading program has a separate number of students, but they all are included in the total number of students in the school's fifth grade (ninety-five students). Because of this, they all fit in a circle.

Florence Nightingale

Florence Nightingale (1820–1910), the nineteenth-century nurse and statistician, used pie charts during the Crimean War to illustrate the death rates of her patients. She is also credited with creating the polar area diagram, a variation of the pie graph.

Labeled, each section within the entire 100 percent circle would look like a pie. The pie is represented by the entire circle and is divided into slices, which each show the number of students reading at a particular level. The chart shows within each slice both the raw number of students and the percentage of the total that represents. The pie effect of all the sections gives a clear visual representation of how each reading level compares with the others.

THE MEAN, THE MEDIAN, AND THE MODE

How to Describe the Center of the Data

As you progress in your study and analysis of data and how to present it in the form of a graph or a chart, there often seems to be a center, or a middle, number where the data groups or concentrates with a high frequency. This central mass of the numbers can be easy enough to estimate when the data is on a graph.

This section will deal with how to determine the center of the data pool and explain why that's important. More specifically we'll look at what is the middle value of the data. Middle values come in two types, the mean and the median.

NUMBERS AND DATA NEAR THE MIDDLE

Remember, statistics are used to mathematically help answer a question. Because statistics and data science can be such powerful tools in their ability to crunch the numbers, you can use the power of statistics to more accurately find the middle of your group of observations. You can also see how far away from the middle each separate observation or data point is.

THE MEAN

Imagine you're going to use statistics to help you find information about a stock you'd like to trade with your online brokerage account. One of the things you are trying to figure out is the average price of the stock when the overall stock market (measured by the S&P 500 index) is up or down by more than 1 percent. You've determined that this is a good experiment to run to help you answer the most important question you have: how does this stock react when the S&P 500 makes a large move up or down? For data you collect the past 100 gains and losses of the stock when the US stock market moved up or down by 1 percent or more.

More Is Better

When collecting data points or collecting bits of information related to answering the question of a study, the more data collected the better! Most professional researchers state that the minimum number of each data section is 100 data points. In other words, for each measurement in the study experiment there should be at least 100 data points collected from at least 100 observations.

The observations of the percentage gains and losses of the stock will be skewed: that is, they'll show a generalized random order. In other words, the stock's gains and losses will range from low to high.

This isn't quite enough information: what you really need to know is on average how much the stock moves when the S&P 500 gains more than 1 percent, and on average how much the stock moves when the S&P 500 loses more than 1 percent.

By determining the mean, you will be able to measure the middle of the data, or the average, of the frequency of both gaining and losing stock prices. By knowing the mean, or the average, gain of the stock on winning days and the average loss of the stock on losing days, you're well on your way to finding the answer to your study's question. In fact, having made this small statistical experiment, you may want to take things further and ask more complex questions about your stock's performance.

Calculating the Mean

Calculating the mean of a group of observations is simple: add up all the amounts in the data points in the set, then divide by the number of data points in that set. This will be the mean and is shown by "x-bar," or $\bar{x}$.

THE MEDIAN OF DATA POINTS IN A GROUP

The median of a group of data points is the point that lies in the middle of the data after the data has been sorted from lowest to highest (or vice versa). For example, if the data points are 12, 14, 17, 21, 23, 24, and 29, the median will be 21 because there are three data points less than 21 and three data points higher than 21. "But," you say, "what if the data points were 12, 14, 17, 21, 23, 24, 29, and 35?" The data points can be formed into two groups, each having the same number of elements. In this case, the median is the mean of the two data points in the middle of the set. That is, the median would be set at $\frac{21 + 23}{2} = 22$.

THE MODE OF DATA POINTS IN A GROUP

The mode of a group of data points is simply the number that occurs the most frequently in the data. While the mean shows the average of the numbers in the observations, the mode shows the data point in the observations that is the most frequently observed. In the stock gain/loss study, the mode is the percentage gain of the stock that shows up the most often. If you discover that gain percentages of both 0.4 percent and 0.75 percent are the most frequently observed, then the entire list of gains of the stock is said to be bimodal; that is, having two modes in the group.

DAMNED LIES AND STATISTICS

British prime minister Benjamin Disraeli reportedly said, "There are three kinds of lies: lies, damned lies, and statistics." That is to say, it's possible to use statistics to support statements that may not be accurate.

Let's look at an example of this from the NBA players' salaries from 2017 (source: www.espn.com/nba/salaries). For the purpose of making this example a little easier to understand, all salaries for the 434 players paid that year are rounded to the nearest million dollars. The mean of the data is $7.1 million, the median is $3.4 million, and the mode is $1 million.

After a group of fans in a bar hear this, a friendly argument breaks out about how well the NBA players are paid. One person claims that they are paid very well, as the average salary is $7.1 million. The next person argues that they really aren't paid all that well because the average salary is only $1 million. The third person argues that the salaries are fair since the average salary is $3.4 million. Who is telling the truth? They all are. People tend to use the word "average" to indicate the mean (it's tempting to say, "It means the mean," but that confuses the issue), but all three measures—the mean, median, and mode—are measures of central, or average, tendency.

Of the three measures of central tendency, the median is actually the best stand-alone measure because it is not affected by the magnitude of the data (as is the mean). The mode is unreliable because of the possibility that there is no one data point that occurs more frequently than any other or that there are multiple data points that occur with highest frequency. For example, suppose that some superstar is signed to a $100 million NBA contract. The mean rises to $7.4 million, the median stays at $3.4 million, and the mode remains at $1 million. (If you wish to argue that the mode did not change in this example, you would certainly be correct. However, if you look at the original data set in which the salaries were not rounded to the nearest million dollars, you will see that the data set has no mode.)

Of course, some of this depends on who's looking at the data. The player who's making $100 million a year probably thinks he's pretty well paid; the player who's making $1 million, perhaps not so much. Again, both viewpoints are valid.

All of this is to say that you have to exercise caution when looking at statistics. They can be deceptive.

THE RANGE AND INTERQUARTILE RANGE

Describing How Wide the Data Is Spread

Once data is collected, we look at its spread. A data grouping's spread is a description of how small its smallest number is and how large its largest number is. Once you know a group's range, you will be able to figure out more statistical information.

A NUMBER GROUP'S RANGE

Let's look at a simple example of finding the range of a number set. After taking a survey, you've collected the following bits of data: –12, –10, –9, –5, –1, 2, 5, 8, 9, 10.

To find the range of the numbers, you first figure out how many places away the largest number is from the smallest number. In this case, –12 is 22 numbers away from the largest number, 10. 10 – (–12) = range of 22 units.

Observations

Remember, in a survey each bit of data collected comes from one individual observation. Each observation can be a question on a survey, a stock price when collecting financial data, or an actual observation from an interpersonal study.

Here is another example with numbers that are not closely arranged around zero:

740; 750; 777; 821; 872; 975; 1,020; 1,040

In this case the range is 1,040 − 740 = 300 units.

What if the same group of numbers also included a much smaller or much larger number within the group? For example, what if the data group looked like the following?

13; 740; 750; 777; 821; 872; 975; 1,020; 1,040

In this case, the range would be 1,040 − 13 = 1,027

THE PROBLEMS USING ONLY RANGE

What is the problem with this example? Well, for one thing, in the first example, the range is 300 units, whereas in the second example, with the addition of only a single number to the data, the range is much larger: it's jumped all the way to 1,027. Although the numbers in both sets are nearly identical, the second set has one far smaller number: 13. In the second set, 13 is so far away in value from the other numbers that it is almost unrelated to them.

Remember, we are imagining that this is a group of data from a survey and that each separate number in the list is from one observation.

In both sets of numbers, most of the numbers are closely grouped together. Because the 13 is very far away from the other numbers, it appears to be a one-time event. When you have a large group of numbers that are loosely centered and one of the data points is far away from the main group, the far-away number is called an *outlier*. Outliers can throw off the range of a group of numbers; because of this you'll need to look at the *interquartile range*.

THE INTERQUARTILE RANGE

A group of numbers' interquartile range (IQR) is the middle 50 percent of the range. In this way, you are measuring a tighter grouping of numbers and therefore are better able to see the true range, since it doesn't include any outliers.

Percentiles

Statisticians regularly rank data. One of the methods of describing the data is to divide it into different sections and identify the section in which the data is located. One of the classic divisions is percentiles. Like percentage, *percentiles* means to divide the data into one hundred sections. When a doctor tells new parents that their child's weight is in the 48th percentile, she is saying that their child's weight is equal to or greater than the weights of 48 percent of children the same age whose weights have been measured over years by the medical profession.

How do you do this? By measuring the middle 50 percent range. First, arrange the numbers in order from smallest to largest. Mark the median number. Next, in the first 50 percent, find that group's median number and mark it. This point is called the *first quartile*. Then do the same with the second 50 percent. This point is called the *third quartile*. The IQR is the difference between the third quartile and the first quartile.

For example, consider the set of numbers:

25, 28, 30, 32, 37, 40, 42, 45, 46, 47, 100

There are 11 pieces of data so the median for the data is the sixth data value, 40. Examine the values that are less than 40 (25, 28, 30, 32, 37). Of these, 28 is the number in the middle, so 28 represents the first quartile. When you examine the data values larger than 40 (42, 45, 46, 47, 100), the middle of these numbers is 46, so 46 is the third quartile. The IQR for this data set is $46 - 28 = 18$. As you can see, the IQR shows only the middle numbers and naturally excludes any outliers.

If a group of numbers is closely centered but has one number far removed, it can throw off the calculated range of that group of numbers. A rule of thumb to help determine whether a piece of data is an outlier is this: the data point is an outlier if it is more than 1.5 times the IQR from the minimum (if the data value is much smaller than the rest of the data) or from the maximum (if the data point is much larger than the rest of the data).

The data values—the minimum, the first quartile, the median, the third quartile, and the maximum—are called the five-number summary, and statisticians use them to create a graph called the box and whisker plot. The box and whisker plot give a nice visualization of the distribution of the data. The following box plot shows the scores for the SAT exam at a small high school:

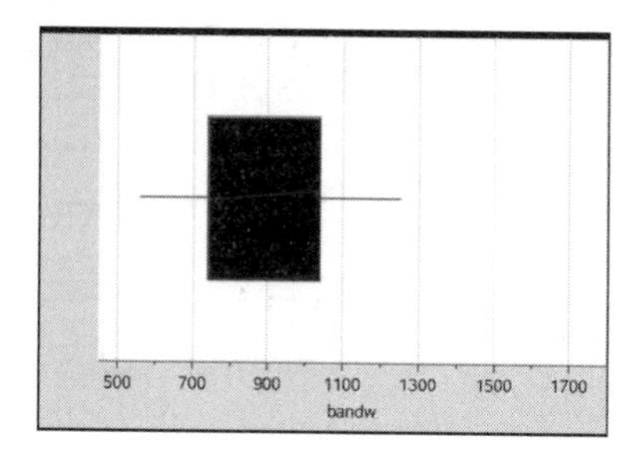

The minimum, first quartile, median, third quartile, and maximum can be easily read from the graph.

Outliers can change the shape of a distribution. The box and whisker plot will show outliers as separate dots from the whiskers of the plot. For example, if 1,600 is added to the data, the box and whisker plot will look like this:

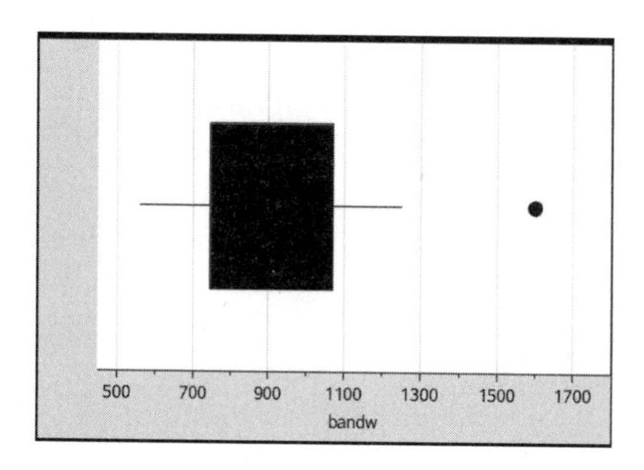

MEAN DEVIATIONS AND VARIATIONS

More Ways to Measure the Shape of Data

Descriptive statistics can be used to explain and describe large groups of numbers that you might have collected from a study, survey, or other information gathered from a database. You can use a number group's variance and deviation to measure how widely the number group is spread out. In other words, variance and deviation tell us how much of the data is spread around the mean and, therefore, how tightly the group of numbers is arranged.

WHAT VARIANCE TELLS YOU ABOUT THE DATA

The mean amount of a soft drink dispensed by a particular brand of machine is 12 ounces. Does that mean each machine will dispense 12 ounces of soft drink every time it is used? No, it does not. The mean amount of liquid dispensed is 12 ounces. The question now becomes: "If the mean is 12 ounces, how many ounces different than 12 are dispensed, and how important is this difference?" One of the measures of the difference is called *variance*. In essence, variance is the sum of the square of the value of the data points and the mean of the data points, divided by the number of data points. (That is, the variance is the mean of the square of the difference of each data point from the mean.) The symbol for the variance of a data set is

s^2 if the data is a sample of the population, and the square of the lowercase Greek letter sigma, σ^2, if the data is the population.

Data: 13, 15, 16, 21, 25

$$\text{Mean: } \mu = \frac{13 + 15 + 16 + 21 + 25}{5} = \frac{90}{5} = 18$$

$$\text{Variance: } = \sigma^2 =$$

$$\frac{(13 - 18)^2 + (15 - 18)^2 + (16 - 18)^2 + (21 - 18)^2 + (25 - 18)^2}{5}$$

$$= \frac{25 + 9 + 4 + 9 + 49}{5} = \frac{96}{5} = 19.2$$

We need to make a slight adjustment about the variance. When computing the variance of the population, the sum of the squares is divided by the number of pieces of data, n. However, when computing the variances for a sample, the sum of the squares is divided by $n - 1$. We do this because of certain desirable properties that make it appropriate for statistical inference. (More will be said about statistical inference later in the book.)

A large group of numbers can be described by how tightly the numbers are gathered around the mean, or average, of the numbers. It's easy to think about: if there are 100 numbers and 80 of those numbers are within 10 percent, either higher or lower, of the average, then it's easy to see that most of the data is centered around the mean. This leaves only 20 percent of the data occupying the remaining 90 percent of the range.

STANDARD DEVIATION IS RELATED TO VARIANCE

The standard deviation of a group of numbers is a way of describing the variance of the group of numbers, but it removes any negative numbers, outliers, or numbers larger or smaller than the average. It is a purer form of describing the number set and can be useful when you need a quicker, though still accurate, way of doing so. In this way, it acts like a smoothing effect, as it takes away some of the extreme outliers from the bell curve.

The standard deviation is calculated by determining the square root of the variance. How does it help to know the standard deviation? Standard deviation can tell the same information as a number group's variance, but it tells it in simpler terms. This doesn't act as a distortion, but rather puts all the information in common terms, so when you go to compare them, you are now comparing apples to apples. For example, if the units associated with the data points are inches, the units of the variance are square inches, while the units of the standard deviation are inches. The symbols for the standard deviation are s and σ, depending on whether the data is a sample or the entire population.

The standard deviation for the set of data 13, 15, 16, 21, 25 is $\sigma = \sqrt{19.2} = 4.38$.

Standard Deviation

While there is a little bit more to the standard deviation than this, it is worth your while to think of the standard deviation as being the average difference between the data values and the mean of the data.

In this way, when you are looking at a large group of numbers spread out far along the bell curve, and these numbers are spread near and far away from the center of the curve, knowing the number group or data group's standard deviation can be very helpful. Why is this so? Mainly because by looking at the standard deviation of the group, you are looking at how far the data is placed from the center. Working with the standard deviation is especially useful when you are comparing data sets to each other: each data set's deviation is on an equal footing.

THE LAW OF LARGE NUMBERS

Probability Measured with Many Tests

In order to get information from statistics, you need to measure a group of data points. If you have a small set of data, the accuracy of some tests will be thrown off—in fact the results can be so far off that they are misleading. The way to combat this problem is to build your tests with very large groups of data. The law of large numbers states that statistical tests are best done with large groups. The best tests are done with groups of 100 or more.

TOO FEW NUMBERS

Much of statistics is a system for testing a set of data. You've learned that the tests measure averages: how far the data lies from the center. Keep in mind that when you use statistics to determine the averages, you can also know how close to the average the data occur most of the time by looking at the shape of the data distribution. If the average happens more frequently, it's considered centered around the mean, and the measurements will be considered to be better quality. Why is this important? For one thing, by studying this you will be able to know what happens most of the time.

Basic Probability

Probability is essentially relative frequency—that is, it is the ratio of the number of times something happens to the number of possible outcomes.

This is why having a large group of data points is important. For example, let's say you are doing market research for a large coffeehouse chain that offers its customers gourmet coffee and premium drinks such as cappuccinos and caramel macchiatos; it also offers free Wi-Fi and premium lunches. One marketing question that the chain's owners may have is "How many of our customers who come in for a regular (relatively inexpensive) coffee could be potential customers for our premium drinks (which can cost up to twice as much as a regular coffee drink)?"

To answer this question, the chain hires your marketing company to survey its customers. You build a questionnaire to ask the customers as they come into the restaurant. You could ask a question with a yes/no answer: "Would you like to try a caramel macchiato today instead of your regular coffee drink order?"

Take a Large Sample for the Study

Because the chain has a worldwide presence, with hundreds of thousands of customers, it is impractical to question every customer who comes in the doors in every location. In this example, the total of all of the chain's customers is called the *population*. From this population, you'll need to take a sample set—say 1 out of 100 customers, or 1 out of 1,000 customers. The number of customers you need to ask needs to be only large enough to get an accurate reading of the entire population. This is called the sample set.

Remember, the sample set needs to be at least 100 samples to make a good and accurate test of the question. While it may be true that the chain has many stores all over the United States, Europe, and Asia, you will get accurate results by surveying approximately 1,000 people from a variety of countries. With simple, binomial surveys, the question can be asked fifty times of random people at twenty different locations. The randomness of the survey can be further enhanced by varying the time of day the question is asked, as well as surveying both foot traffic and drive-through traffic.

All this is important because if the group that takes the survey is too much the same, it will skew your answers. Also, if the group is all from the same area, it might skew your answers culturally, since people from that area may drink only one kind of drink. Finally, if you don't ask enough people you won't get a wide enough sample of what the customers really think: ten, twenty, or fifty people could be random, but there still is a high chance that the sample set will be off and the numbers won't read as true.

EMPIRICAL PROBABILITY AND SUBJECTIVE PROBABILITY

Probability with Multiple Sets

Probability is the measure of whether and how frequently something will happen. This section will show how statisticians calculate the probability that things will happen with multiple sets, as well as cover some of the basic laws of probability.

PROBABILITY BASED ON PAST EXPERIENCES

Subjective probability refers to a situation when you are thinking, "What are the chances something will happen?" and you base your answer on what you've experienced in the past. Let's imagine you are shopping at your favorite antiques store. You're an avid antiques collector, and you've shopped the antique malls of your hometown for years. You also know that the antique mall booth owners fill up their booths with fresh antiques every Sunday, when the shops are closed. Because of this information, and your past experiences, you know that the best times to find good deals are Monday mornings.

While you've not collected data or made a survey, you do have years of experience behind you, and you can therefore make a statement: "There is a high probability that I will find good antiques at the antique malls on Monday morning." Subjective probability is characterized by loosely based statements and few to no hard numbers.

When you are using subjective probability, you are using your experiences to make an educated guess about expected outcomes. This contrasts with another form of probability that measures probable outcomes using numbers and statistical measurements. This is called *empirical probability.*

EMPIRICAL PROBABILITY

We've learned that the probability of an event happening can be measured by a survey or a sampling of the entire population. Statistical analysis of probability is governed by the following rules:

Rules of Probability

- The chance of an event happening *must* be between 0 percent and 100 percent (between 0 and 1).
- If the chance that an event happening is x, then the chance that it doesn't happen is $1 - x$. If the probability that it will rain today is 40 percent, then the probability it will not rain today is 60 percent.
- If the chance of an event happening is 1, then it will happen 100 percent of the time. The probability that the sun will rise in the east tomorrow is 1.
- If the chance of an event happening is 0, then it will never happen. The probability that the sun will rise in the west tomorrow is 0.
- In a group, each event probability must add up to 1 (100 percent). When rolling a fair die (singular of dice), the outcome can be 1, 2, 3, 4, 5, or 6. The probability of getting a 1 is written P(1). $P(1) + P(2) + P(3) + P(4) + P(5) + P(6) = 1$.

- If there are two separate events that can happen together, this is called a *compound event*. When drawing a card from a well-shuffled deck of bridge cards, drawing the ace of spades can be considered a compound event because it is both an ace and a spade.
- If there is an event that happens in either/or two separate groups, this is called a *union of events*. For example: there is a sale today at either antique store A, antique store B, or both antique stores A and B.
- If an event happens in both A and B it is called an *intersection of events*. For example, both antique store A and antique store B have antique pocket watches for sale.
- If an event is either A or B but not both, it is called a *mutually exclusive event*. For instance, a registered voter is either a Democrat or a Republican. No one can be registered in two parties simultaneously.
- An event that is outside all that is contained in set A is called a *complement of an event*. An example is sales that occur at antique stores or non-antique stores. Those that occur at non-antique stores are a complement of an event.

CONDITIONAL PROBABILITY

There is another way to measure probability. This one involves answering the question "What are the chances that something will happen if something else has already happened?" It assumes that event A is dependent upon event B. This more complex form of probability is called *conditional probability*. It can become very complicated to figure out, since many events can trigger a second event. For example, pick a card from a well-shuffled deck of bridge cards, look at it, put it back in the deck, and draw a second card. What is the probability the second card is an ace? The answer is 1/13 (there are four aces in the 52-card deck). The drawing of the first card had no impact on the selection of the second card. Repeat the experiment, but this time don't put the first card back. What is the probability now that the second card is an ace? This time it depends on whether or not the first card was an ace. If it was, the probability that the second card is an ace is 3/51 (one of the four aces has been removed from the deck), but if the first card was not an ace, the probability the second card is an ace is 4/51.

Independent Events

Two events are said to be independent of each other if the occurrence of one event has no impact on the probability the second event will occur.

Let's take another example: the results of a survey completed by 782 people are displayed in the accompanying Venn diagram.

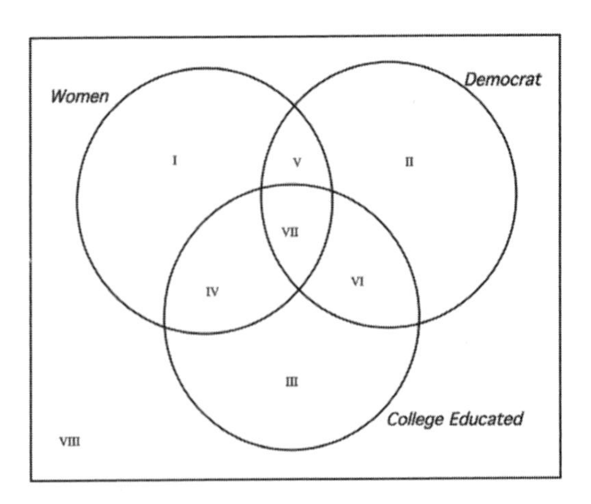

The three key characteristics of the diagram are gender, political affiliation, and education.

- Region I represents women who are neither college educated nor see themselves as Democrats.
- Region II represents men who are not college educated and see themselves as Democrats.
- Region III represents college-educated men who do not see themselves as Democrats.
- Region IV represents college-educated women who do not see themselves as Democrats.
- Region V represents non-college-educated women who see themselves as Democrats.
- Region VI represents college-educated men who see themselves as Democrats.
- Region VII represents college-educated women who see themselves as Democrats.
- Region VIII represents non-college-educated men who do not see themselves as Democrats.

Before we ask questions about probability based on these survey results, let's take a moment to analyze the eight regions in the Venn diagram.

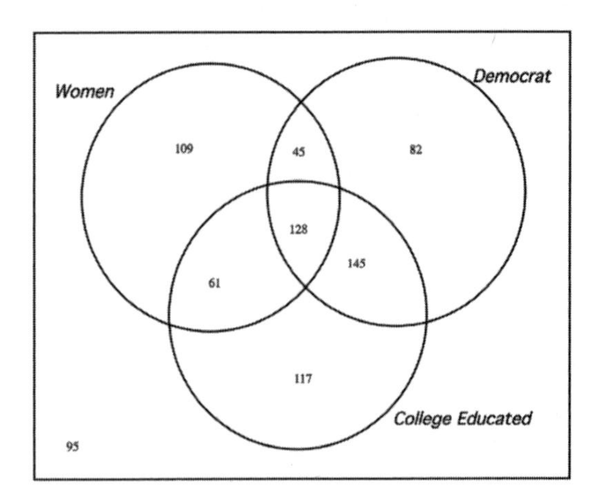

You have to be careful when reading a Venn diagram. Entries that are not contained inside the circle labeled "Women" are men—there is no other choice. Entries outside the circle labeled "Democrats" can have any of a number of political affiliations—Republican, Independent, Green Party, Socialists, etc. Outside of the "College Educated" circle are those people who did not attend college. Did they attend but not finish college? Or did they never attend at all? We don't know. Did they finish high school? Again, we don't know. All we do know is that they classified themselves as not college educated.

Here are some questions for us to consider to help with reading the Venn diagram.

If a person from this survey is selected at random, what is the probability that:

1. The person is a woman? Of the 782 people surveyed, 343 (109 + 61 + 128 + 45) are women, so the probability is 343/782.

2. The person is a college-educated woman? Of the respondents, 189 (61 + 128) are women with a college education, so the probability is 189/782.
3. The person is a woman or is college educated? There are 343 women, plus 262 (117 + 145) men who claim to be college educated, so the probability is 605/782.
4. The person is college educated given that the person is a woman? "Given that the person is a woman" tells us to concentrate only on the 343 women in the survey (all contained in the circle called "Women." Of these women, 189 are college educated, so the probability is 189/343.
5. The person is a woman given that the person is college educated? This time we limit our discussion to the 451 college-educated people from the survey. We just showed that there are 189 women with a college education, so the probability is 189/451.
6. The person is a Democrat given that the person is a college-educated woman? Of the 189 college-educated women, 128 claim to be Democrats, so the probability is 128/189.

Probability in the Financial Industry

Probability is commonly used in financial applications. Hundreds or thousands of events can trigger changes in the price of stocks, gold, or other commodities. They can affect economic announcements, binomial stock events, etc. To examine all the triggers, sophisticated computer models are used to analyze vast amounts of data. This builds the probability model. From there, complex mathematical formulas can give the data analyst answers. The questions may be as simple as "What will be the direction of the S&P 500 index in the next one minute? Up or down?"

YES OR NO

Binomial (or Bernoulli) Distributions

The seventeenth and eighteenth centuries were a fertile period of mathematical discoveries. Breakthroughs in the understanding of probability, calculus, and geometry occurred in England and on mainland Europe. The animosity that was so prevalent during that era (and is still alive today) among the various governments also existed among the scientific and mathematical communities. Most popular among these is Gottfried Leibniz's (1646–1716) publication of the principles of calculus before Isaac Newton (1642–1727) was able to publish his *Philosophiae Naturalis Principia Mathematica* (in which Newton published his work on mathematics and physics). One of the major players in this battle was Jacob Bernoulli (1654–1705), a Swiss mathematician who studied under Leibniz. Among his many accomplishments, Bernoulli developed the process of the binomial (or Bernoulli) distribution.

YES OR NO OBSERVATIONS

Most of the time when you are conducting or building a survey, collecting data from the Internet, or researching data from external sources, there are many, many different variations in the quantity and range of numbers (data) in the information you've collected. However, sometimes you are making observations of data regarding questions that can only be answered with a yes or a no answer.

Additionally, most of the time each yes/no question has no relation to the other yes/no questions asked in the survey. In other words, they are *independent* of each other. If you answer question 1 with yes, then you can still answer question 2 with yes or no—it's not dependent upon how question 1 was answered.

Because they are independent of each other, the answers can be recorded as only yes or no. Also, each different time the question is asked again, it can be answered without any connection to any other time it was previously answered. This is a simple way to explain binomials.

For example, prior to the interview for determining a donor's health history, the donor is asked to read some educational material. The first question on the health history is "Are you feeling well today?" and the second question is "Did you read the educational material?" The donor may answer yes to both questions during one donation and then no and yes on the next donation. The questions and the events are independent of each other.

Moriarty and the Binomial Theorem

Binomials were given lasting fame by Arthur Conan Doyle (1859–1930). In his story "The Adventure of the Final Problem," the great detective Sherlock Holmes is giving Dr. Watson a vivid sketch of his archenemy, Professor James Moriarty. Moriarty is, Holmes tells the good doctor, "endowed by nature with a phenomenal mathematical faculty. At the age of twenty-one he wrote a treatise upon the binomial theorem, which has had a European vogue." To be fair, in the pastiche *The Seven-Per-Cent Solution*, by Nicholas Meyer (1945–), Watson asks Moriarty (who in this version is no criminal but merely Holmes's old mathematics tutor) if he's written anything on the binomial theorem. "Certainly not," Moriarty exclaims. "Who has anything new to say about the Binomial Theorem at this late date?"

How are these used? Suppose a woman is hired to work at a new company or start-up and is offered a salary plus stock options. These stock options are often binomial in nature. The contract states that if the new employee works for a certain number of years at the new company, she will receive a flat monthly salary. In addition to this salary over the years, there will be an equity option. The equity option is to the effect that if the employee is still employed within five years of hiring, and if the equity stock of the company exceeds a certain dollar amount, then the employee will have the "option" of exercising her right to own a certain number of shares at that price.

With this type of contract, the employee is better off if she stays for the full five years, and she is also motivated to work hard and help the company earn as much as it can, so that its stock value meets or exceeds the value of the exercise price of her stock options.

You can see from this description that the contract is binomial: in five years, exercising the options can be formulated as a yes or no choice: yes, if the price of the stock exceeds the contract option price, or no, if the price of the stock never got high enough for the employee to exchange her options for real stock.

Because the contract is yes/no, it can be worth only one of two values: a preset amount or $0. In other words, the contract is either yes (the full dollar amount of the contract) or no (worth $0).

Each contract that is written for each employee is usually written at a different exercise price and a different amount of stock options for each employee. The CEO generally gets the best deals; other employees get contracts that would require that the stock price rise to a higher value before the options were worth anything.

BUILDING MODELS WITH BINOMIALS

When building a spreadsheet of the data you've collected for your statistical study, normally you record each type of data or each question asked in your discovery; each one goes in a separate column on a spreadsheet. Each observation point (person surveyed, trading day, classroom time observed, etc.) is in the horizontal row of the spreadsheet. In that way, each observation point has all the separate points (questions, data points, measurements, etc.) in columns along the same row. You'll want to keep the numbers of these in the same scale, to make it easy when you compare the numbers later.

In the case of a question in your observations that requires a yes or no answer rather than a number, let's convert the yes or no answers into numbers. This makes it easier to compare like data. In this case the yes answers are converted to 1, and the no answers are converted to 0. Changing the answers to ones and zeroes also helps when performing other, higher-level statistical analysis—for instance, multiple regression analysis. We'll discuss this kind of analysis in a later section.

BINOMIAL DISTRIBUTION

Imagine taking a test consisting of twenty true or false questions and answers. Having not studied for the test, you have to guess at every question. If the passing grade is 65 percent, what is the probability that you pass the test? In other words, what is the probability of getting at least 13 questions correct?

Using some of the language from probability and statistics, we need to collect twenty observations in which the probability of a correct response for each observation is 50 percent and each event is independent of every other event. This constitutes the essence of a binomial experiment. There are three conditions that must be met in a binomial (Bernoulli) experiment:

- There can be only two outcomes per trial—call them success and failure
- There must be *n* repeated, independent trials
- The probability of success in each trial must be constant

In our example, there are two outcomes per trial—true or false. There are twenty repeated independent trials. The probability of a correct answer in each trial is 50 percent (since you did not study and are presumably guessing on each question). The Microsoft Excel function BINOMDIST is designed to compute binomial probabilities. Entering the pertinent information for this problem, you will discover there is only a 13 percent chance of passing the test. Our advice is that you study for the test.

Jacob Bernoulli

Jacob Bernoulli (1655–1705) was one of four members of his family who studied mathematics, physics, and engineering during the seventeenth and eighteenth centuries. The other three are renowned in their own right.

Is rolling a die an example of a binomial distribution? Is selecting a card from a well-shuffled deck? The answer is simple: no and yes. Let's look at the problem of rolling a fair die. For the sake of argument, you want to see how many times you get a 6 when you roll the die 100 times. First roll: did you get a 6? Here is the critical step. If your answer is "No, I got a 3," then your experiment is not binomial. If your answer is "No, I did not get a 6," then you do have a binomial experiment. Remember: there can be only two outcomes per trial. Success—you got a 6. Failure—you did not get a 6. (In the binomial version of the experiment, there is no need to know what number the die displayed if it was not a 6.)

A similar argument can be made for drawing a card from a well-shuffled deck. Did you get a spade? Yes or no. But there is also a second issue: did you put the card back in the deck before drawing the next card? If you did, you have a binomial experiment because the probabilities remain the same as the first trial. If you did not, then you do not have a binomial experiment because you've changed the probability that the next card drawn can be a spade.

CENTER AND SPREAD

The distribution of the number of successes for n independent trials has a center and a spread that are both measureable. If the probability of success on any one trial is p, the mean of the distribution of successes is the product np, and the standard deviation for the distribution is $\sqrt{np(1-p)}$. The mean is also called the expected value, so we get the notation $\mu_x = E(x) = np$, while the standard deviation has the notation *ox2*. For example, if a company who makes cold calls claims a probability of making a sale on any call is 0.05, then an employee who makes 500 cold calls in a day is expected to make $(0.05)(500) = 25$ sales with a standard deviation of $\sqrt{500(0.05)(0.95)} = 4.87$.

BINOMIAL VERSUS NORMAL DISTRIBUTION

Technically speaking, these are two very different situations. The binomial distribution uses a discrete variable. We count the number of successes gained from the *n* trials. The normal distribution uses continuous variables. We measure the thickness of the washers. However, when we look at the distribution of the probabilities of *x* successes in 20 trials of a process for which the probability of success on any one trial is 0.5, the distribution has a very familiar shape:

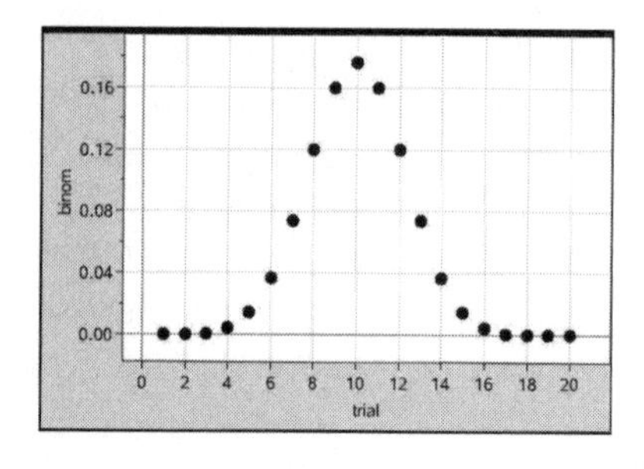

When the probability of success is 0.4, we can see a little bit of skewing to the left:

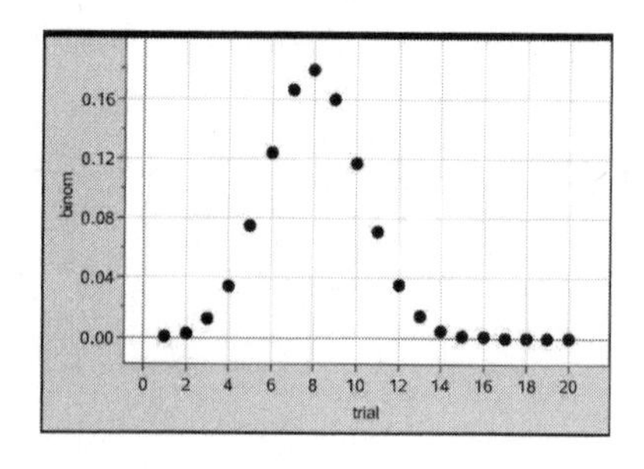

Is there a relationship between the two that can be used? Most definitely. As we saw, when the probability of success is different from 0.5, there will be some skewing of the symmetry of the distribution. Research has shown that if n is sufficiently large, the skewing can be offset. What is sufficiently large? If the values np and $np(1 - p)$ are each larger than 5, the sample size is sufficiently large.

But what about the discrete versus continuous variable issue? The probability of $n = 4$ with a continuous random variable is 0. If we have a binomial experiment and we are seeking to determine the probability of 4 successes with a normal approximation, we set the bounds of the normal distribution from 3.5 to 4.5 (which clearly includes 4 and no other possible discrete outcomes).

An extension of the binomial distribution is the Poisson distribution. In a Poisson distribution, we seek to determine the probability for a discrete number of successes, but we also include a continuous interval. Some examples of a Poisson distribution include the number of calls received per hour, the number of cars served at a tollbooth per day, and the number of customers who enter a bank per hour.

BASICS OF PROBABILITY DISTRIBUTIONS

What Are the Chances?

As we've seen, statistics can help you answer the question "What are the chances?"—that is, how frequently is something going to happen? This is a good question to ask and answer if you are trying to predict an outcome or if you are trying to measure how often something will happen. Probability is one of the key ideas in modern statistics, especially in the realm of statistical finance. This section will discuss how to determine "What are the chances?"

THE RANDOMNESS OF THINGS

While it seems in life that many things are boring, mundane, and predictable, the truth is that much of life must deal with things that are random. The traffic jam that you encounter going to work, the spring thunderstorm that soaks you while walking to the train, the event in overseas news that pushes the stock market lower today, and the numbers that come up in the lottery that make your $2 lottery ticket worthless—all of these are examples of randomness.

We usually think of positive things as predictable and mundane, and we think of things as random when they turn negative or against our best hopes. The facts are, most individual events in life have a large element of randomness attached to them. Can you predict with 100 percent certainty that you will make it into work on time, that the same people will be your clients, that you will take the same way home at night, or that the same people will be at the coffee shop for your afternoon break? You can make a very reasonable assumption, based on probability, that all will be the same, but there will be exceptions: for instance, an unexpected accident on your route to work that blocks traffic for half an hour and makes you late to the office. The point here is not to anticipate all possible individual outcomes; it's to find the most accurate and efficient means of determining the *probable* outcome.

The Irrationality of Observations

Just because something is improbable doesn't mean we don't somehow expect it to happen. The outstanding example of this is the lottery. You know that you've played the Powerball lottery every Saturday night for the past ten years. You play only one ticket, for $2. You've collected data that includes 520 observations (10 years × 52 Saturdays a year), and you've seen that your lottery tickets have never won the lottery. In fact, you also know from comments in the newspapers and online that the true chances of winning the lottery are close to 1 in 290 million. But every Saturday you still go to the store to pick up your lottery ticket. Why? Because some part of your mind believes that you can be an outlier and beat the odds.

The entire industry of gambling and the city of Las Vegas have been built on the foundation of this irrationality.

Every day you do just that: make predictions based upon past observations. You know that all summer, Thursdays are pretty much the same: get to the office, do your work, meet with clients—right down to your same afternoon coffee at the same coffee shop and the same family trip to the baseball diamond with your children. Because this same sequence has happened every Thursday all summer, when a new Thursday comes, you're going to make a reasonable assumption that all will be the same as previous times. Why? Because you are using statistical methods to say to yourself, "The average things that happen to me on Thursdays are this, this, and this....I've made this observation many times (each Thursday this summer), and therefore, things that happen on Thursdays won't stray far from the mean (i.e., everyone and everything act pretty much the same as every average Thursday)."

As you can see, chances can be very strong that something will happen the same way each time, or they can be equally strong that an event will never happen. While this may be true, there is still an element of uncertainty, because as we know from our discussion of bell curves, there are always outliers. Let's look at what could happen differently:

WHAT ARE THE CHANCES THIS THURSDAY ISN'T THE SAME AS LAST THURSDAY?	
Chance of rain in July	4 days out of 31
Chance of traffic jam in LA	102 hours/year/typical driver in traffic jams
Chance of you becoming sick	Approximately 1 day every 3 months as a sick day
Chance of power failure	1 hour/year
Chance of police chase on the freeway	1 day every 5 years

THE CHANCES ARE—THE ODDS ARE...

You can see from the chart that the chances of the five different things on each Thursday not happening have different odds. Even though you know from experience that the chances of this Thursday being pretty much the same as the last Thursdays are very high, you see here that each one of the day's elements has a chance of not happening properly. In other words, the odds must be "beaten" on each element every time for the day to have a different event. You have to have no rain (and therefore no canceled baseball games), you also have to beat the odds of a traffic jam (so you have time to get your coffee before the game starts), then you have to beat the odds of not having to take a sick day (and therefore not going to work in the first place), and so on. To find the odds of the day happening in the same way as other Thursdays, you have to consider everything that could go wrong (or differently) and take this into consideration!

This is statistically what your mind is thinking when you say to your child, "Be ready at 5:30 tonight for the baseball game. I'll be home to pick you up and we'll go from there."

ANALYZING PROBABILITY DISTRIBUTIONS

The Truth about Odds

When you are predicting how your workday will play out, you are using your past knowledge of how workdays have gone in the past. You know that in the morning, there is the commute and your usual stop for a coffee and a roll at the same corner donut shop, you know that most of the time you'll have a full tank of gas (because you usually fill up every Saturday morning), and you also know what will happen during work (especially if you've been working there for a long time). You may also know what happens on the way home since you make the same commute as always: the kids, dinner, taking the kids to their baseball or soccer games, a quiet evening, and bed. You also know the same thing will most likely happen tomorrow.

How do you know this? Because your mind is making small, educated calculations of what will most likely happen. These are statistical assumptions based upon many years of observed data.

WHAT IS REALLY GOING ON?

Statistically what is really going on behind the scenes is that almost every time this type of day starts, the same types of things happen. In other words, you make (subconsciously) the same observations. It is as if your mind is taking a statistical survey of what will happen: each event during the day is a bit of data in the survey; each time it happens (the same or differently) is an observation.

Since most of the time the same things are happening day in and day out, your observations form an imaginary data set of an average day. You've learned that when you make many observations in a study and you put all the observations under a bell curve, the center of the bell curve will be the average thing that will happen. Therefore, the average day that has happened in the past is the center of the bell curve. Anything else that has happened that is substantially different will be located away from the center; the less often an event occurs, the farther away from the center it will be. Most of the times the average day happens, most of the data will be in the center of the curve, so it will form a tall bell curve. This is because there is so much data around the middle and so little away from the middle. From this, you can draw some conclusions as to the probability distribution.

A tall bell curve says that most of the time the average will happen. You can then say, "The probability of some odd, weird thing happening today is very small, because my bell curve says that the probability distribution is very centered around the mean." If it rained and you were late for work, those events would be a small amount away from the average day. Suppose you went for your morning coffee and roll, and the shop was out of baked goods. That event doesn't happen too often, and it wouldn't be typical. If everything was average during your workday, but you came home and the coach had canceled your child's soccer game, that would also not be typical. But it's not so far off the typical and is still very close to an average day.

From the shape of the curve, you would be able to say that the probability of the same day happening is high: the probability distribution is closely centered around the mean, or closely centered around the average, typical day.

RANDOMNESS IN STATISTICAL TERMS

When we are discussing events that are random, the classic way to explain them is with a coin toss. The experiment is simple: toss a heads/tails coin four times. What are the different ways the group of coin tosses come up?

COIN TOSS		
Toss 1	Heads	Tails
Toss 2	Heads	Tails
Toss 3	Heads	Tails
Toss 4	Heads	Tails

As you can see, with each coin toss there are two possible outcomes: either heads or tails. There is a 50/50 chance of each happening with the first coin toss. What are the chances of four heads in a row coming up if the coin is tossed four times? The chance of the heads in the first toss is 50 percent, and then the chance of the next toss is also 50 percent. This means that half of the first time and half of the next time the coin will come up heads, or 50 percent × 50 percent = 25 percent. You could say with certainty that the chances of two heads in a row of two tosses are 1 in 4, or 25 percent. To continue your winning trend of flipping heads, you would have to up the odds to 50 percent for the first toss, 50 percent for the second toss, and 50 percent for the third toss: 50 percent × 50 percent × 50 percent = 12.5 percent, or a 1 in 8 chance that you will win your three heads in a row coin toss. As you can see, your odds are getting worse as the number of coin tosses increases.

Who uses this information in real life? Casinos, for one. They can figure out the odds of a number coming up at the roulette table. If the odds of the number are 50:1, then they know, on average, the gambler will lose his bet fifty times for every one win. This is an average and is compiled over many games over many weeks, months, and years. If the odds are 1 in 50 that the casino will have to pay out, then the casino will make the payout on that number less than 50 times the bet. This way, no matter how many times the game is played, the casino will earn the *differential* of the two bets. If they pay out 10 times and they earn 50 times, then the casino will earn a steady 40 times per bet. Make sense? Think of it this way: the casino runs the table so that when they pay out, which they almost never do, they pay out way less than they take in (which they do almost all the time). In the casinos, the house never loses!

Measuring the risk of going to the tables in a casino is the same as measuring the probabilities of the outcome that something will happen.

THE ROLL OF THE DICE

Statistical Ways of Showing Chance

The basic way to figure out how often something will happen is the same as finding out its probability. Probabilities are best figured out mathematically by watching how many times something happens during a test run of random events. This section will show you how to use statistics and math shorthand to show the chances of things happening during a random test sample.

THE SECRET TO THE DICE

The toss of a die is an excellent way of learning how to statistically analyze probability. This is important, since probability is used in many aspects of modern medicine, finance, and marketing. When a new cancer drug is being tested, the drug undergoes an extensive testing procedure. This procedure is designed to measure how often the drug will have effects upon people who take it. At the same time, in marketing studies, surveys are taken to find out how often people will use a website to buy a product. In this case a study is made showing the website to a large group of people. Software is installed that follows the click-through path that the shopper takes through the site and ultimately keeps track of how many times the customers buy something from the site. In finance, the probability that stock prices will go up when bonds go down is often measured. This data is used when professional portfolio managers invest.

Dice have six sides. If the dice are normal, there is an equal chance that any number 1 through 6 will be rolled. There is a 1 in 6 chance that a 1 will show up for every die thrown.

ACCURATELY MEASURING PROBABILITY

If you throw a die and you get a 1, your chances are 1 in 6. If you throw the die two more times, the chance of getting a 1 a second time (that is, two times out of three) is small, but there is a chance this will happen. If indeed you rolled a 1 twice out of three rolls, you might assume (by the initial look of the facts) that the chances of rolling a 1 twice in three rolls is 2 out of 3.

The Probability of Compound Events

The probability of rolling a 1 with a single die is 1/6. The probability that the second roll of the die results in an outcome of 1 is also 1/6 because the two events are independent of each other. Therefore, the probability that the first two rolls each result with a 1 is $1/6 \times 1/6 = 1/36$.

Why is this misleading? Because you've only rolled the die three times! The chance of rolling the die and getting 1 two times out of three rolls is small—about 1 in 200, to be exact. The only realistic conclusion that you can make is that you've beaten the odds with your three rolls.

Rolling the die only three times doesn't really tell you much. In fact, it doesn't really help you at all when it comes to proving your theory that you have a 1 in 6 chance of rolling a 1 when you roll the die.

How can you overcome this huge error? Well, first you'll need to increase the number of times you roll the die and record the number of times a 1 comes up. In fact, to get accurate measurements of how often you'll get a 1, you'll need to roll the die a minimum of 100 times. Most professional researchers know that running a random test 100 times is the bare minimum to get any results that matter. It would be better to run the "roll the die" test 500 or 1,000 times and record how often a 1 came up. It's only after you run a very large number of tests that you can draw a conclusion of "What are the chances" or "What is the probability" that the 1 will happen.

THE SAMPLE SET AND THE OBSERVATIONS

In the example of rolling the die, there are two main factors that are used to statistically describe the probability of getting a 1. Let's say you've rolled the die 1,000 times and seen the 1 come up 166 times. The 1,000 rolls are equivalent to the number of times you've tested your theory of the 1 in 6 outcomes. These 1,000 rolls are called the sample set. In statistics, the sample set is the big picture, the entire group in your test or in your study. If you are measuring the effectiveness of a new cold medicine, you might decide to test 1,000 students on a college campus. If it's a large public college, there might be too many students to test every student. This certainly would be the case for some very large universities. In this case you could take a *sample* of the students at random and test how well the cold medicine is working. You could randomly sample students in the hallways to find a well-diversified study group: this is called the *sample set*.

You would then ask if they took the cold medicine and how well it was working. Your test would be somewhat scientific and random enough to be effective.

The Beginning of Probability

The correspondence between Blaise Pascal (1623–1662) and Pierre de Fermat (1601–1665) concerning a problem posed by Antoine Gombaud, Chevalier de Méré (1607–1684) in 1654 laid the fundamental groundwork of probability theory.

In the case of the die roll, the 1,000 rolls would be your sample set. Surely you wouldn't be able to observe every roll of every die in the world forever, but with 1,000 rolls at random you'll get a good sampling of the typical die roll. With both the cold medicine and the die roll sample set, what are seen are called the *observations*.

NORMAL DISTRIBUTION

Area under the Bell Curve

Up to this point, we have looked at probability for discrete events. We now take a look at a very special case for probabilities of continuous data, the normal distribution (also known as the Gaussian or Laplace–Gauss distribution). The graph of the normal distribution is the bell curve. The mean, median, and mode of a normal distribution are all the same point, and the bell curve is symmetric about this point. The standard normal distribution has a mean equal to 0 and a standard deviation equal to 1.

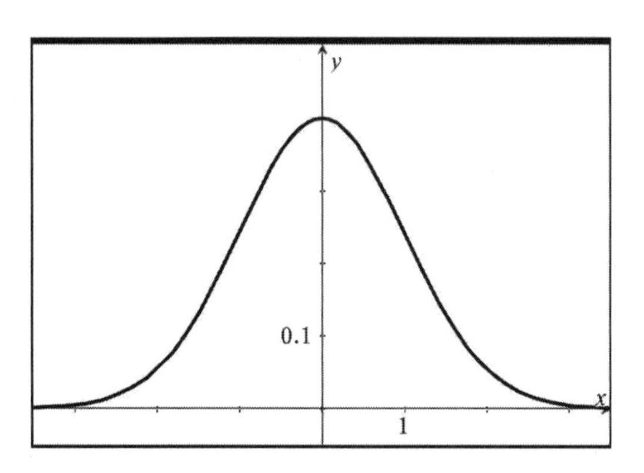

PROPERTIES OF THE BELL CURVE

It can be shown, using calculus, that the total area under the bell curve is 1. While the reality is that few distributions are actually normally distributed, many distributions approximate the normal distribution. The fact that the total area is equal to 1 enables the distribution to be used as a probability model for applications. In the next section, we will look at the central limit theorem, which makes the use of the normal distribution all that more practical.

The histogram shown below is the relative frequency distribution for the thickness of 1,500 washers used in a commercial process. Observe how the histogram looks like the bell curve.

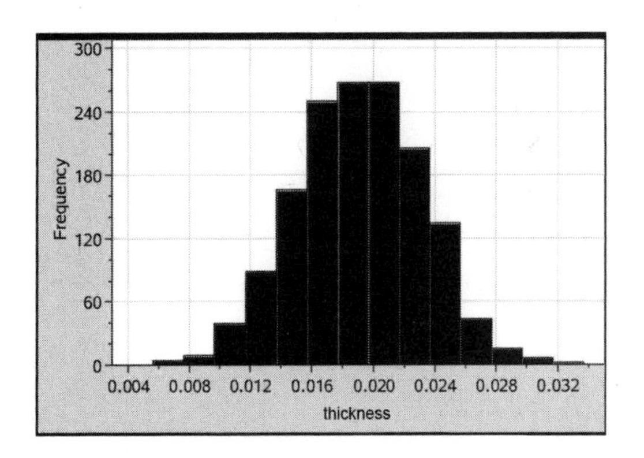

The table displays the relative frequencies for each of the intervals.

RELATIVE FREQUENCIES FOR THICKNESS (IN CM) OF 1,500 WASHERS		
INTERVAL	FREQUENCY	RELATIVE FREQUENCY (%)
0.0057–0.0077	4	0.2
0.0077–0.0097	9	0.6
0.0097–0.0117	39	2.6
0.0117–0.0137	88	5.9
0.0137–0.0157	166	11.1
0.0157–0.0177	250	16.7
0.0177–0.0197	268	17.9

RELATIVE FREQUENCIES FOR THICKNESS (IN CM) OF 1,500 WASHERS		
INTERVAL	FREQUENCY	RELATIVE FREQUENCY (%)
0.0197–0.0217	268	17.9
0.0217–0.0237	205	13.7
0.0237–0.0257	134	8.9
0.0257–0.0277	44	2.9
0.0277–0.0297	16	1.1
0.0297–0.0317	7	0.1
0.0317–0.0337	2	0.4

If a washer from this manufacturing process is selected at random, we can state that the probability that the thickness of the washer is between 0.0217 cm and 0.0237 cm is approximately 13.7 percent.

EMPIRICAL RULE

There are certain relationships that are true for the area under the bell curve.

- The area to the right of the mean is 50 percent
- The area within 1 standard deviation of the mean (that is, $\mu - \sigma < x < \mu + \sigma$) is approximately 68 percent
- The area within 2 standard deviations of the mean (that is, $\mu - 2\sigma < x < \mu + 2\sigma$) is approximately 95 percent
- The area within 3 standard deviations of the mean (that is, $\mu - 3\sigma < x < \mu + 3\sigma$) is approximately 99.5 percent

(The first item should be obvious, and the last three can be shown with calculus.)

For example, suppose the thickness of the washers highlighted in the last section is normally distributed with a mean of 0.2 cm and a standard deviation of 0.015 cm. What is the probability that a washer selected at random will have a thickness between 0.185 and 0.215 cm? The answer is 68 percent because this is the interval within one standard deviation of the mean. Between 0.17 and 0.23 cm? The answer is 95 percent because this is the interval within two standard deviations of the mean. Between 0.155 and 0.245 cm? The answer is 99.5 percent because this is the interval within three standard deviations of the mean.

What if the thickness of the randomly selected washer is between 0.17 and 0.215 cm? This is a bit trickier. The interval from 0.17 cm to 0.2 cm is 47.5 percent of the data (that is, half of 95 percent), while the interval from 0.2 cm to 2.15 cm is 34 percent (half of 68 percent). Therefore, the result is 81.5 percent.

And if the washer's thickness is greater than 0.215 cm? We know that the probability that the thickness is greater than 0.2 cm is 0.5 and that the probability that the thickness is between 0.2 and 0.215 is 0.34. The probability that the thickness is greater than 0.215 must be $0.5 - 0.34 = 0.16$.

What about between 0.18 and 0.22 cm? The empirical rule does not help with this problem. The correct probability (81.76 percent) can be found with the use of appropriate technology.

Chebyshev's Theorem

While the empirical rule applies to data that are normally distributed, Pafnuty Chebyshev (1821–1894) showed that for any distribution, at least $1 - \frac{1}{k^2}$ of the data will lie within k standard deviations of the mean. For example, at least $\frac{3}{4}$ of the data will lie within 2 standard deviations of the mean.

STANDARDIZED SCORES (AKA Z-SCORES)

In the days before technology, one used a table to compute probabilities from a normal distribution. Since there are an infinite number of possible distributions with different means and standard deviations, the standardized normal distribution was designed to measure the number of standard deviations a raw score was from the mean. If x represents a raw score, then the z-score is computed by the formula $z = \frac{x - \mu}{\sigma}$ (for populations) or $z = \frac{x - \overline{x}}{s}$ (for samples). Examine these two problems:

1. The heights of the students at Central High School are normally distributed with a mean of 68 inches and a standard deviation of 2.4 inches. What is the probability that the height of a student chosen at random from the student body at Central High School is between 64.4 and 74 inches?
2. The weights of the chassis for the G style automobile made by a high-end manufacturer are normally distributed with a mean of 1,000 pounds and a standard deviation of 12 pounds. If a chassis for the G style automobile is randomly selected from the production line, what is the probability that the chassis weighs between 982 and 1,030 pounds?

The answers to these questions are the same because they both represent the intervals from 1.5 standard deviations below the mean to 2.5 standard deviations above the mean.

$$(z = \frac{64.4 - 68}{2.4} = -1.5; z = \frac{982 - 1000}{12} = -1.5;$$

$$z = \frac{74 - 68}{2.4} = 2; z = \frac{1030 - 1000}{12} = 2)$$

Fortunately, the built-in functions in modern technology allow us to enter the raw data values to compute the probabilities.

NORMAL SHAPES OF BELL CURVES

As you can see from the table for the thickness of washers, most of the time data comes within one or two standard deviations from the mean. In some very rare cases (a 1 in 333 chance), a bit of data will fall beyond three standard deviations. How does this apply in the real world?

Let's take the stock market, where professional traders and money managers pick investments such as stocks, commodities, gold, and bonds after making careful studies of how these investments have reacted in the past to varying market conditions. These professionals work to build a basket of investments (called a portfolio) that are designed to achieve one goal: to earn the greatest amount of return with the minimum of risk. They use statistics to build models of how different combinations of investments react under different conditions (good and bad markets). They then take out and add different stocks, commodities, bonds, etc., to try to get the highest return they can. They know that when some investments go down, others trend up. If they build the right model with statistics, they can build the right portfolio of the best blend of assets.

They look at past historical data of how different investments have done under different up and down economic and stock market conditions. From a statistical analysis of this data, they build a model of the most likely way the investments will react in the future. In this way they are using the empirical rule: they know 95 percent of the time, the investment's price will be within two standard deviations from the average. Because of this they can pretty much predict where the stock or bond will be in the future.

The problem with this theory is the very small chance that the stock will act unusually. In fact, there is only a 0.3 percent chance the stock's price will be farther than three standard deviations from the average price of the stock. Professionals know the chances of this happening are so small that they act as if it will never happen—but in fact it does. Investment portfolio modelers were caught off guard by the rapid fall of the stock market (and their investors' portfolios) during the 2007–2008 worldwide financial crisis.

The investment models didn't fail; it is just that the 0.3 percent chance event happened.

THE CENTRAL LIMIT THEOREM

Sampling Distributions of the Mean

As we discussed in the previous section, there are many distributions that approximate the normal distribution. We begin with sampling from populations that are normally distributed and then look at distributions that are not normally distributed. In this section, we discuss what might very well be the most important theorem in all of statistics: the central limit theorem. It is this theorem that allows us to make educated guesses about population means and population proportions.

SAMPLING FROM NORMAL POPULATIONS

Suppose we have a population that we know is normally distributed with a mean equal to μ and a standard deviation equal to σ and that we draw unlimited samples of size *n* from the population. (This means we can draw a sample of size *n* and then "put all the pieces back" before we draw the next sample.) We create a new set of data by computing the mean for each sample drawn. What is the mean, $\mu_{\bar{x}}$, and standard deviation, $\sigma_{\bar{x}}$, for this distribution of sample means? The mean of the distribution of sample means will be exactly the same as the mean of the population. That is, $\mu_{\bar{x}} = \mu$. The standard deviation for the distribution of sample means, called the *standard error of the mean*, has the formula $\sigma_{\bar{x}} = \dfrac{\sigma}{\sqrt{n}}$. An immediate impact is to see that the values of n get larger the standard error of the mean gets smaller. Graphically, this will cause the bell curve to become much narrower and taller.

As an example, suppose a manufacturer ships goods in a box so that the weights of the boxes are normally distributed with a mean weight of 25 pounds with a standard deviation of 4 pounds.

1. What is the probability that a box picked at random will have a weight between 24 and 27 pounds?

We'll leave the answer in terms of z-scores for the purpose of making a point. The z-score for 24 pounds is ($z = \frac{24 - 25}{4}$ = -0.25 and the z-score of 27 pounds is $z = \frac{27 - 25}{4}$ = -0.5. So we are looking for the probability P(-.25 < z < 0.5).

2. What is the probability that the mean weight of a sample of 25 boxes will be between 24 and 27 pounds?

The z-scores for these values are now $z = \frac{24 - 25}{\frac{4}{\sqrt{25}}}$ = -1.25 and $z = \frac{27 - 25}{\frac{4}{\sqrt{25}}}$ = -2.5. Even without being able to compute the actual probability, you can see that P(-1.25 < z < 2.5) will be significantly larger than P(-.25 < z < 0.5).

3. What is the probability that the mean weight of a sample of 50 boxes will be between 24 and 27 pounds?

The z-scores for these values are now $z = \frac{24 - 25}{\frac{4}{\sqrt{25}}}$ = -1.25 and $z = \frac{27 - 25}{\frac{4}{\sqrt{25}}}$ = -2.5.

Can you see that by taking larger sample sizes, you can be assured that the probability of getting the correct sample mean is increased?

THE CENTRAL LIMIT THEOREM

The normal distribution is a powerful tool to compute probabilities for continuous random data—but not all continuous random data necessarily follow a normal distribution. Then again, we aren't too often as concerned about a single piece of random data as we are about the mean of this data. Pafnuty Chebyshev (1821–1894) and a number of other Russian mathematicians from the Saint Petersburg Imperial University observed a powerful behavior of data that takes advantage of the normal distribution. It is called the central limit theorem. The theorem states that as the size of each sample gets large enough, the sampling distribution of the mean can be approximated by the normal distribution no matter what the distribution of the individual data might be. With this theorem, statisticians are able to gain more information about populations. Consider the population of 2,000 randomly selected integers between 0 and 100,000 whose distribution is displayed in this graph:

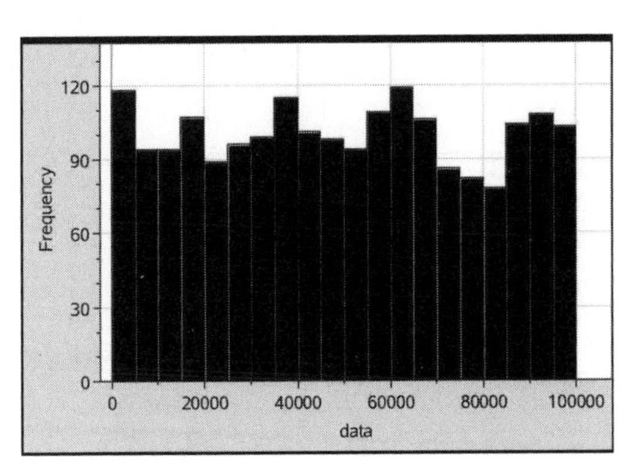

One thousand random samples are taken from this population. First the size of each sample is 10 elements. Then the size of each sample is increased to 25, then 30, 100, and 500. The mean of each sample is computed and stored. The distribution of the sample means is shown. (Please observe that the scale of the window needs to be changed as the sample sizes get large.)

For a sample size of 10, the distribution looks like this:

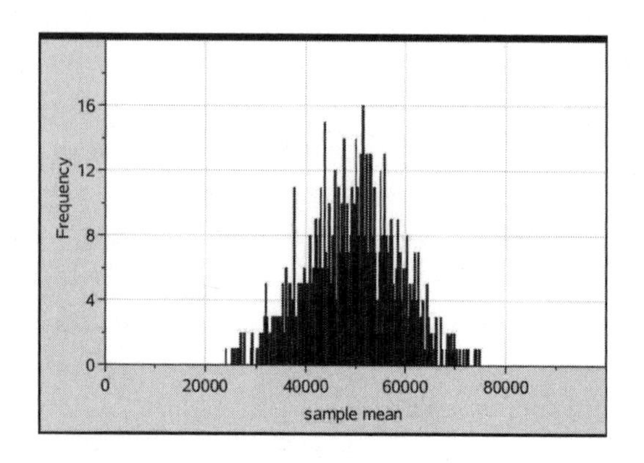

For 25, it looks like this:

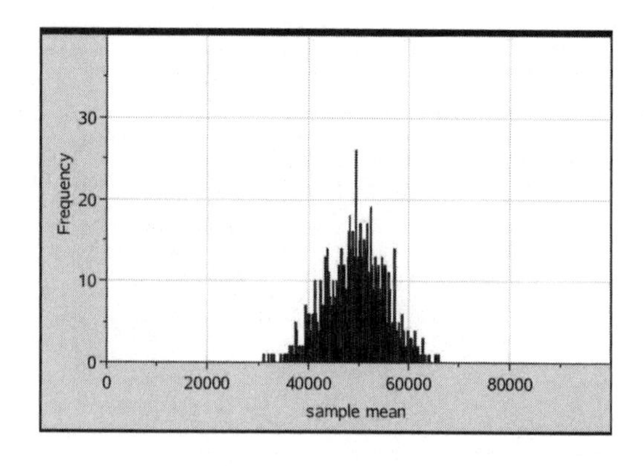

For 30:

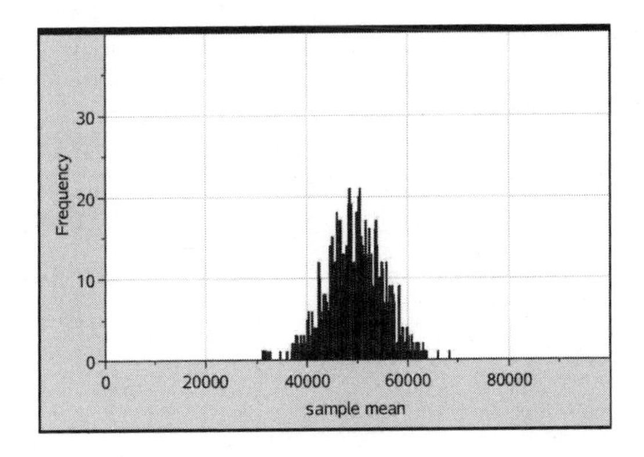

For 100:

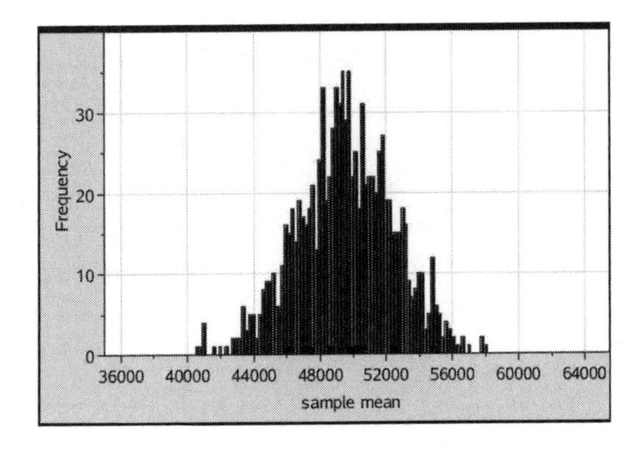

And finally, for 500:

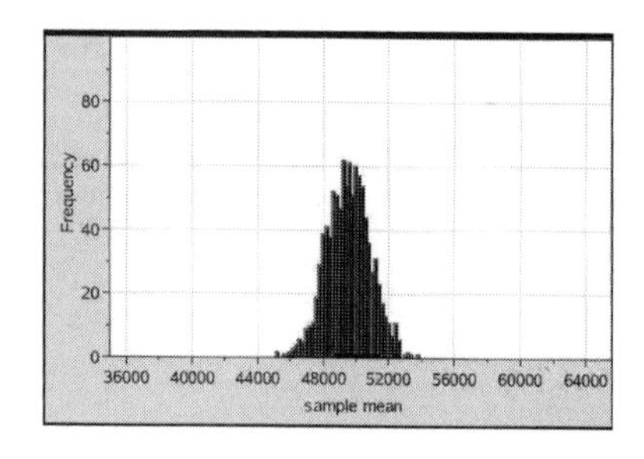

SAMPLING DISTRIBUTIONS OF THE PROPORTION

There are times when the parameter (the population value rather than the statistic of the sample) of interest is the proportion of the population rather than the mean. Once such instance is the percent of the population who prefer candidate X for the presidency. Lower case p is used to represent the sample statistic, while the lower case Greek letter *rho*, ρ, represents the population parameter, and σ_ρ represents the standard error for the proportion.

$$\left(\sigma_\rho = \sqrt{\frac{p(1-p)}{n}}\right)$$

When can we apply the normal distribution to the sampling of the proportion? Given a sample size n and a proportion of the sample p, we require that np and $n(1-p)$ each be larger than 5.

Here is an example applying the sample distribution of the proportion. A car dealership manager has determined that 30 percent of the customers who purchased or leased a new car from her store are repeat customers. She draws a random sample of 150 customers who have purchased or leased a new car from her dealership. What is the probability that between 30 percent and 35 percent of these customers is a repeat customer?

Can we apply the normal distribution to this problem? Yes, because $(0.30)(150) = 45$ and $(0.70)(150) = 105$. The z-score for 30 percent is 0, while the z-score of 35 percent is $z = \frac{.35 - .30}{\sqrt{\frac{(.30)(.70)}{150}}} = \frac{.05}{\sqrt{\frac{.21}{150}}} = 1.336$. All we need to do is apply our technology to determine $P(0 < z < 1.336) = 0.4092$.

OUTLIERS ON THE BELL CURVE

Fat Tails and Whale Tails

As we learned earlier, when looking at a bell curve, there are one, two, and three standard deviations. With normal events, it's pretty safe to assume that 95 percent of the time an event or observation will take place within the range of two standard deviations away from the average.

Standard deviations give a sort of assurance, a loose-fitting guarantee that things are predictable and that they most likely happen how they are expected to occur. While this is true most of the time, occasionally something happens that isn't normal. These events are beyond the norm and sometimes can be very disruptive to otherwise very well–laid plans.

PAYING THE PRICE FOR PREDICTABILITY

In many cases, statistics can provide a very nice comfort level about what is most likely to happen. Occasionally, though, they can mislead. One of the most famous recent cases of this is the 2008 recession and stock market crisis. Before this event, many professional money managers were using statistics to build well–thought-out investment portfolios. The methods they were using went back to concepts that had been around for more than fifty years. The concepts of diversification, "Don't put all your eggs in one basket," and portfolio insurance were based on layer upon layer of statistical theory. This information told the money managers that their money was safe; it was invested properly; and if all else failed, the idea of diversification would save the day—and their portfolios.

The statistics they were using showed that most of the time, stocks, bonds, and other investments acted a certain way (going up or down) depending on market conditions. They did historical research and built statistical models to support their strategy. They knew, through their data collection, sample sets, and observations, that investments followed the laws of statistics: 95 percent of the time things would happen as predicted. If the 95 percent happened within a range of up/down make money/lose money, they could buffer against the losses. Such investments were called "good."

THE OTHER 5 PERCENT

The money managers knew that statistics worked. These were mathematical laws that were proven. The only problem was that the economic conditions in the United States, Europe, and Asia were anomalous: the 5 percent events started to happen. Most important, the housing bubble in the United States abruptly burst. Housing prices plummeted, and homeowners were stuck with houses that were worth less than they owed the banks. Banks, some of which had been selling toxic mortgages to people who couldn't afford them, suddenly discovered they were holding worthless paper.

The stock market reeled. Stocks that had in the past moved only \$2–\$5 up or down on a given day began moving \$10–\$20 up or down. Violent shifts shook the markets. The Dow Jones Industrial Average plunged almost a thousand points. Then it began to steadily drop, with sudden short upward spurts. As one commentator put it, "The Dow is falling with an occasional bounce off a window sill."

Things got so bad and so many of the 5 percent things happened that a new term was coined: a *black swan event*. The thought was, the stock market events were so rare, that they had the same chances of happening as a black swan being born. These "far away from the average" events began happening 15 percent, 20 percent, or even 30 percent of the time. Financial chaos, for a time, became the new normal.

Black Swan Events

The statistician and investment analyst Nassim Nicholas Taleb (1960–) developed a metaphor called the black swan theory that seeks to explain how extremely rare anomalies occur and why we have a hard time predicting them and often try to rationalize them in hindsight. His argument is essentially that we're blinded by psychological prejudices that always lead us to expect things will perform in a usual way. When they don't, we often argue that we *could* have foreseen the black swan event, but for one reason or another, sufficient data wasn't available to us.

Another term that came into being to describe anomalous data is *whale tail*. This is the shape of the bell curve when the outliers begin to dominate. Now, rather than being shaped like a bell, it has the head and tail of a whale, with large amounts of data points at the very extreme ends of the scale.

OUTLIERS ON THE BELL CURVE

While there always is a chance an extreme event will happen, these events are usually infrequent: about 5 percent of the time. It's a statistical fact that in normal circumstances 5 percent of the time a stock price will be far away from its average. How much is "far away"? Well, it's different for each stock. For some, far is $10. For others, it's $1. It depends how that stock has moved in the past. If the money managers researched the price of the stock going back years and discovered that the stock drops $10 or more only 5 percent of the time, then they will build their models with these facts. If they've built the model to follow statistical law, then most of the time it will be a reliable tool. But during the years of the 2008 crisis, conditions were not normal. The outliers became the new norm.

LIMITED AND UNLIMITED DATA

Discrete Random Variables

When the number of data observations is limited, or controlled within only a certain, preset range, the observations are called *discrete random variables*. This section will introduce and describe how discrete random variables are used in statistics.

SAMPLE SETS—DISCRETELY SIZED

The size of data sets varies. The number of observations can be very large, numbering in the thousands, hundreds of thousands, or even millions. To help make the process of looking at the data set easier, an analyst can take a sample of the bigger set. This sample set should be taken at random to get the most effective widespread numbers and thus the most realistic reflection of the entire population of the group.

If the data collected in the sample set can only be within a range of numbers, it is called *discrete* data. For example, say a marketing firm conducts a survey concerning used car dealerships such as CarMax. CarMax sells used cars of all makes. The company knows that at any time they will have customers wanting different types and brands of cars. A certain percentage of the shoppers will be truck buyers, and of these a smaller group will only be Ford truck buyers. CarMax would like to know, "What is the average number of Ford trucks on our lots at any given time?"

CarMax knows that certain cars sell the most, and because they track this, they can try to keep these cars in stock. When they take the survey of what percentage of cars on the lots are Ford trucks, they take an inventory with all their lots throughout the United States. If they know that 30 percent of their customers want trucks, and half of these prefer to buy a Ford, they will try to stock 15 percent of their inventory with Ford trucks.

Discrete and Continuous Data Sets

The usage of the terms in the context of the size of the data sets is different from the context of whether the data is counted or measured.

The total number of trucks that they could possibly have is equal to the total number of spaces in their lots. If they have 1,000 lots, with 250 spaces for vehicles per lot, they can have a maximum of 250,000 Ford trucks (i.e., every space in every lot would be filled by a Ford truck). The minimum number of trucks is 0 (i.e., there would be no Ford trucks at all in any lot). Thus, the possible number of Ford trucks in the CarMax lots has clear high and low limits, which will be reflected in the data set that CarMax analyzes. This is an excellent example of discrete random variables.

CONTINUOUS RANDOM VARIABLES

On the other hand, while the number of trucks that could possibly be on the lot can only range from 0—250,000, the number of customers coming into the car dealership and asking for a Ford pickup truck can vary from 0 to the size of the population who can buy a truck (i.e., of legal age with a license and insurance). On one end, there could be no customers asking for Ford trucks during the period when data is being collected. On the other end, there is an unknown maximum of how many people can come into the dealers across the country to ask for the trucks. Here, CarMax may have to rely on its experience to estimate how many people come into the dealer every week looking for Ford trucks. There is really no way to control how many people come into the dealership. Theoretically, the number could be in the millions (if the dealer offered an incentive, such as free carnival rides, hot dogs, balloons, coffee, handout gifts, etc.). The largest number of visitors to the dealership is unknown and not in a set range. Therefore, this set is called a *continuous random variable*.

Again, these are examples of variables that are measured. Because a discrete data set is finite, it has definite maximum and minimum values that can be included in your model. On the other hand, remember that observations that can be very elastic or have a very large range are called continuous random variables. These are terms that are good to know when you are thinking of what you are measuring.

Statistics are an essential part of the work of stock traders and analysts. They look at the movement of groups of stocks within the market and determine the frequency with which that movement occurs and in what direction. Based on this, they can predict which stocks to buy and which to sell.

Photo Credit © Getty Images/phongphan5922

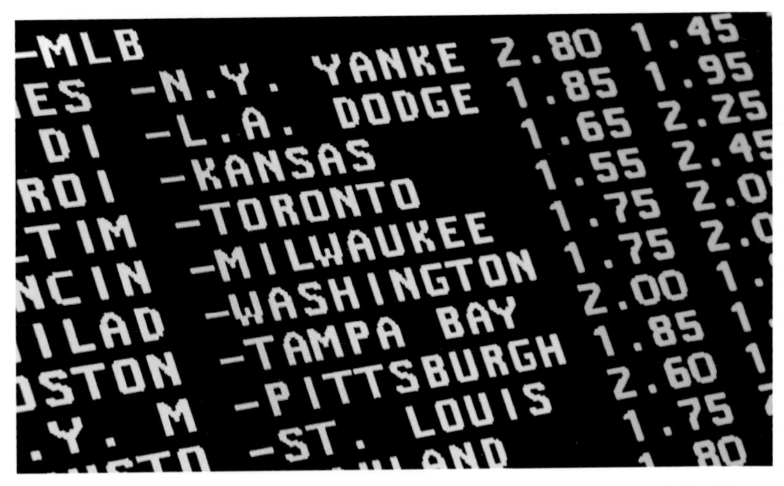

A branch of statistics, sabermetrics, is devoted to the study of baseball statistics. These statistics are used both by teams to decide which players they should try to acquire and by bookmakers in Las Vegas and elsewhere, who use the statistics to calculate the odds (shown here) of teams winning individual games, the playoffs, and the World Series.

Photo Credit © Getty Images/sb-borg

Towns and cities often use statistics to figure out hours and funding for municipal institutions. For instance, a library might poll a representative selection of users to decide what hours it should be open and which parts of the library patrons use the most.

Photo Credit © Getty Images/kali9

Some statistical surveys use multiple-choice questions, offering survey takers a variety of possible answers. Once analysts go over the answers, they can use them to build a model that predicts outcomes. This is the whole object of statistics: to be able to predict things.

Photo Credit © Getty Images/atakan

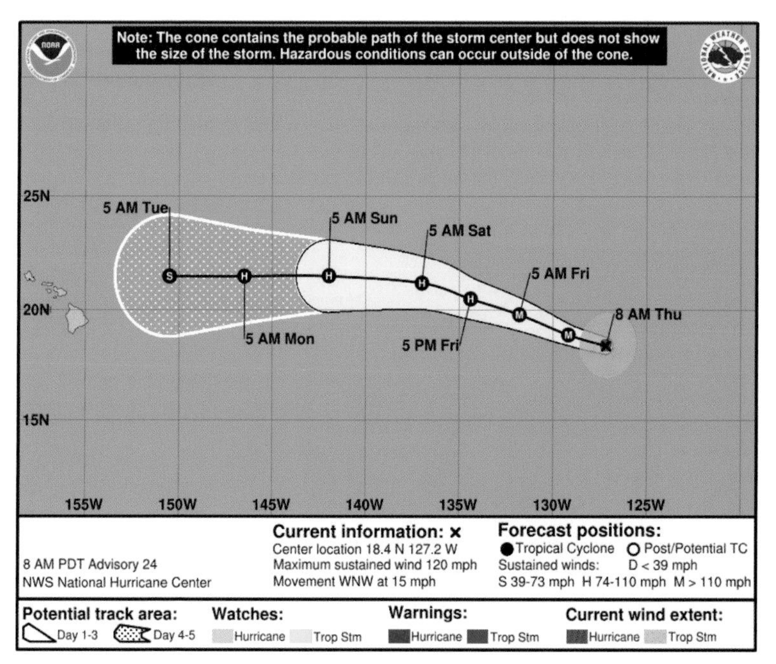

Weather forecasters have vast amounts of data to analyze, but they don't have sufficient time to do so before making a forecast. Instead, they must select a smaller sample that accurately reflects the larger population of data. Forecasters examine previous weather patterns and, based on a statistical analysis, can predict the probability of weather-related events.

Photo Credit © National Weather Service/Public Domain

City and state governments often commission statistical studies of traffic patterns. Such surveys are a useful measure of the wear and tear on the highways, given the average number of cars, small trucks, and semitrailers that pass a section of road in a given time.

In addition to her groundbreaking nursing work during the Crimean War, Florence Nightingale (1820–1910), a gifted mathematician, made several innovations in the field of statistics. She popularized the use of the pie chart and other visual aids to show statistical conclusions.

VARIANCE AS A MEASURE OF RISK

Standard Deviation and Variance of Discrete Random Variables

The shape of a bell curve tells you a lot about what is being studied. In this section, you'll learn about how bell curves can be narrow, wide, or in between. Each one of these shapes tells a different story, and sometimes this story is about risk.

LOOK AT THE SHAPE

Bell curves, as we said earlier, are made from collecting data and plotting that data around the average. First, you plot each bit of data and find the average: values higher than the average go on the right side of the center, and values lower than the average are on the left side of the center. It is normal for bell curves to have most of the data near the average, simply because that's the average!

There can be many different shapes to the bell curve, ranging from very tall and narrow to very flat and wide. The shape of the graph is directly tied to how close the bulk of the data lies to the average. This shape is called the *spread of the bell curve*, or the *spread of the data*. By studying the shape, you can learn about the variance of the data.

The variance of discrete random variables is the average spread of each data point around the mean (the average) of the group. The variance tells about how far from the average observances can vary.

In other words, if you are looking at two identical subjects (like investment performance of two different stocks) and each one returned 10 percent on average per year, you might assume that each stock is equivalent in performance. Looking deeper, you would use statistics to find the variance of each stock's price over the year. If one stock returned 10 percent on your investment, but its daily price went up and down –5 percent to +5 percent, and the other stock returned 10 percent in a year, but its daily price went up and down –2 percent to +2 percent, then the 2 percent stock had less variance. To earn the 10 percent return on the first stock, you have to tolerate a 10 percent daily swing (–5 percent to +5 percent), but the second stock can earn the 10 percent yearly return with only a 4 percent daily swing. The second stock offers the same 10 percent performance but with considerably less risk. In statistical terms, its variance is much smaller.

The variance in this example measures how much of the daily stock price can be expected to move away from the average each time. If we were to plot the performance of the second stock on a bell curve, the curve would be high and narrow, showing a tight spread. Most of the time, the stock's price (the data) is tightly centered around the mean.

The Limits of Statistics

Many investors argue that statistical analysis of the stock market has significant limits. This is partly because the market has so many moving parts and because many of these parts are fundamentally irrational. The 2007–2008 financial crisis, for example, was partly sparked by the fact that homeowners and lending institutions suddenly realized that many homes simply weren't worth what they were currently valued at. No amount of statistical analysis could have predicted that. According to many investors, statistics can tell you quite a bit about the past but relatively little about the future.

STANDARD DEVIATIONS

You've learned that a group can be arranged in a bell curve. You've also learned that the bell curve can have many shapes, and that a bell curve that is high and narrow means that the variance of the data is small and closely set around the average. Also, you've learned that if the bell curve is low and widely set, this means that the variance of the data is large and far set around the mean. If you know the variance of a group of data, then you can also get a better picture of the risk. In the stock example, we saw two stocks with two different variances.

If you are comparing large amounts of variances from a large number of bell curves, then looking at the standard deviation will show all the numbers on an even keel. This can be an effective tool when comparing large groups of bell curves. With stock picking, you can measure which stock offers the greatest return with the minimum risk.

How would you do this? You'd arrange the stocks by standard deviation, with the lowest standard deviation at the top of the list. At the same time you'd look for the stock that historically had offered the greatest return. If your goal was a 10 percent return, you would look for the 10 percent return with the lowest standard deviation. Since these numbers are a measure of the stock's price swing during the year, a stock with the lower price swing but the same return would be considered to be a safer stock and therefore be a better choice for risk/return payoff.

SIZE MATTERS

Samples, Sample Size, and Sampling Distributions

As we've said, when creating a representative sample from a larger body of data, the sample size should be larger rather than smaller. Statisticians usually aim for more than 100 observations, but the rule of thumb is the more the better. With this in mind, is it okay to study a group of less than 100? The answer is yes, but the sample is less random and therefore less reliable.

Gertrude Cox

One of the most well-known figures in modern statistical theory is Gertrude Cox (1900–1978). She studied experimental statistics and founded a department of that name at North Carolina State University. She was the first woman elected to the International Statistical Institute, and in 1956 she was elected president of the American Statistical Association. Some refer to her as the First Lady of Statistics.

When deciding what type of sampling method to use, it is best to keep in mind that the sample must reflect how the entire population of data looks. If the sample isn't random, there is the chance that the information will be skewed, not reliable, or not true.

SAMPLING DISTRIBUTION

When you've created a sampling distribution and collected data, your first task is to take the data and plot it in a bell curve. Then you measure the distribution. Whereas a distribution tells you about the individual data points, sampling distribution gives information about the samples seen as a whole. With this method you can gain the same information, but it has the effect of *information smoothing.* Because the samples are randomly chosen from the larger population, the information and data have become smoothed (*smoothed* meaning more generalized and therefore easier to interpret) through being randomly chosen. While a properly performed population sampling will be varied, diverse, and random, there is still the effect of the result having a rounding, generalization, or smoothing effect. Because of this, the sampling distribution, too, will be smoothed. This smoothing can add to the quality of the picture of the curve, much like a standard deviation uses only the positive square root of the variance.

MEASURING DISTRIBUTION

What If There's No Pattern?

Up to now, we've been assuming that once you determine the form of your analysis, collect data, determine that it's good, and plot it, you can study its shape and learn from it. But what if it has no shape?

If the data looks like it has no pattern (especially on the scatter-plot graph), then the question you're trying to answer can't be found by collecting this type of data. There is no statistically significant pattern to the observations you've collected.

This can be discouraging. When it happens, either the questionnaire or the data set is bad, and you need to rerun the tests with changes. You'll do this by changing the database, looking for other factors to test, or by changing the questions in the questionnaire.

Remember, your data plot should form a bell curve. This is because most of the information is just that: it's "most of the information" and is therefore average in nature.

An Inverted Bell

Sometimes, when you're plotting your data, you'll find yourself looking at an inverted bell curve. The data seems to divide into two extremes, both of which pull the edges away from the center. Many teachers experience this in their grading: large parts of the class get high grades (As and Bs), while equally high parts get Ds or Fs. There are relatively few Cs to prop up the middle of the curve. Even though it's not necessarily the pattern you want to see, it's still a pattern, and you can analyze it.

GOOD DATA HELPS PREDICT FUTURE OUTCOMES

The Excel charting function is one of the best tools to help you determine the quality of your data. The scatterplot chart and the bar chart can show the distribution of the data. All of these are steps on the way to the bell curve, which is what you are looking for. If the chart has no shape, then your data is off. If a bell curve isn't perfect, or a curve is somewhat a bell but is lopsided to one side: that's okay too! Even a bell curve with fat tails, i.e., a bell curve that has a high center as well as somewhat higher ends is okay. The problem arises when the data is all over the place with no pattern.

Remember, you've got a question that you're trying to answer with statistics. If you can't find a pattern in your answer, then something's wrong, either with the question, the data, or your plotting and analysis. Why? Well, in part because we are looking for a repeatable study. If there is a pattern to the data observations, you will be able to take this repeatable pattern to the next level and build a model that would help you predict outcome in the future. That's what modern statistics, data analysis, and data modeling are all about: using past data to help predict future outcomes.

WHAT ARE CONFIDENCE INTERVALS?

Further Tests of Quality

Confidence intervals are used to get an estimate for a population parameter—usually the mean or population proportion. Measuring the quality of your data can be broken down further into point intervals and confidence intervals. A *point interval* is simply the statistic you get from a sample. While it is true that you might get lucky and the sample mean (or proportion) is exactly the same as that for the population, that result is highly unlikely. Rather, we try to create an interval estimate for the value of the population parameter. We take advantage of the normal distribution in doing so because we know that if we take a large enough sample, the central limit theorem guarantees us that the distribution of the sample statistic will be normally distributed, and the empirical rule tells us that 99.5 percent of the data (in this case, the sample mean/proportion) will lie within three standard deviations of the mean/proportion. Since the mean of the sample means is the same as the population mean, we get a decent estimate for the population mean/proportion. (For ease of reading, we'll use either the mean or proportion in our discussion, but please note that the process applies to both parameters.)

MEASURES OF QUALITY OF POLLS

It is common for news stations and newspapers to report the news as what they've learned from polls. They might report news such as "67 percent of New Yorkers hate the subways" or "73 percent of Chicagoans never take a summer vacation." While such headlines can be true, and while the stories that accompany them can be newsworthy or even entertaining, there is an element missing. These stories need to include the confidence interval that the polls were taken with. Confidence intervals can be expressed in the polls as "73 percent of Chicagoans never take a summer vacation (with a 5 percent margin of error)." The *margin of error* refers to the *confidence interval* of the poll.

How is the confidence interval determined? Let's work with a 5 percent margin of error. We are claiming that we are 95 percent certain that our answer is correct (or we are allowing for a 5 percent error).

How large is that confidence interval? Is a 5 percent margin of error good enough? Is it necessary for the sake of accuracy to have a 2 percent margin of error instead? Or even a 1 percent margin of error? What if there was a 10 percent margin of error? Would this tell you the data collected in the poll was good? How can you determine the quality of the data collected in a survey or poll? Easy! You look at the confidence interval.

How is the confidence interval determined? Let's work with a 5 percent margin of error. We are claiming that we are 95 percent certain that our answer is correct (or we are allowing for a 5 percent error). We look for the values left and right of the sample proportion so that 95 percent of the area under the graph lies between these two values.

Associated z-Scores

The associated z-scores for computing margins of errors are: 10 percent — 1.65; 5 percent — 1.96; 1 percent 2.58.

The confidence interval is then computed as $p \pm \text{ME} = p \pm pz\sigma_p = p \pm z\sqrt{\frac{p(1-p)}{n}}$. In the case of the Chicagoans who never take a vacation, let's assume for the sake of this exercise, that 1,000 people are surveyed. The confidence interval for their result is $.73 \pm 1.96\sqrt{\frac{.73(.27)}{1{,}000}}$ giving the interval from 70.25 percent to 75.75 percent.

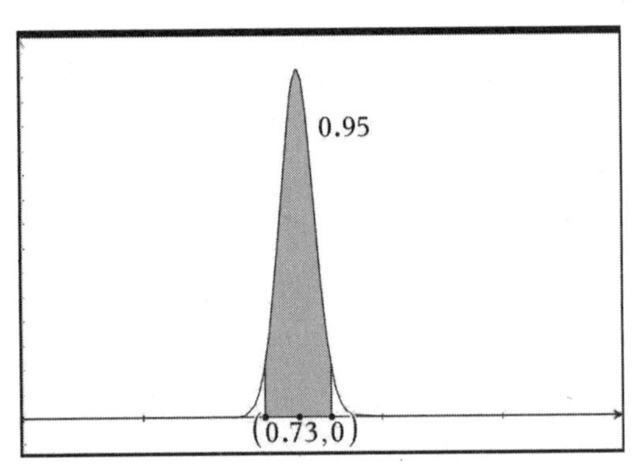

THE 90–95–99 FACTOR

In the real world, confidence intervals of any measure are broken down into 90 percent, 95 percent, and 99 percent confidence intervals. Other percentages are considered, but these three are the

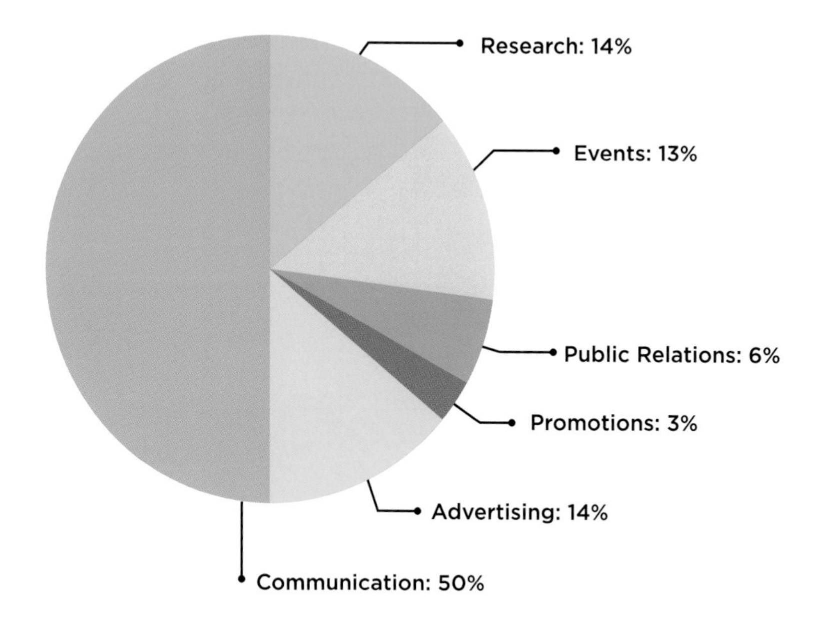

A pie chart shows the proportions of a statistical population that are made up of subgroups. The population percentages add up to 100 percent. Pie charts are used when you want to show the relationship of data points to one another. In this one, which shows the budget allocation for a marketing program, marketers can tell at a glance what proportions of the program are intended for which activities.

Photo Credit © Simon & Schuster/Katrina Machado

Janet L. Norwood (1923–2015) was the first woman to head the US Bureau of Labor Statistics, the department that gathers data on economics and labor and provides it to the government. She was also a leading member of the American Statistical Institute and the International Statistical Institute.

Photo Credit © AmstatNews

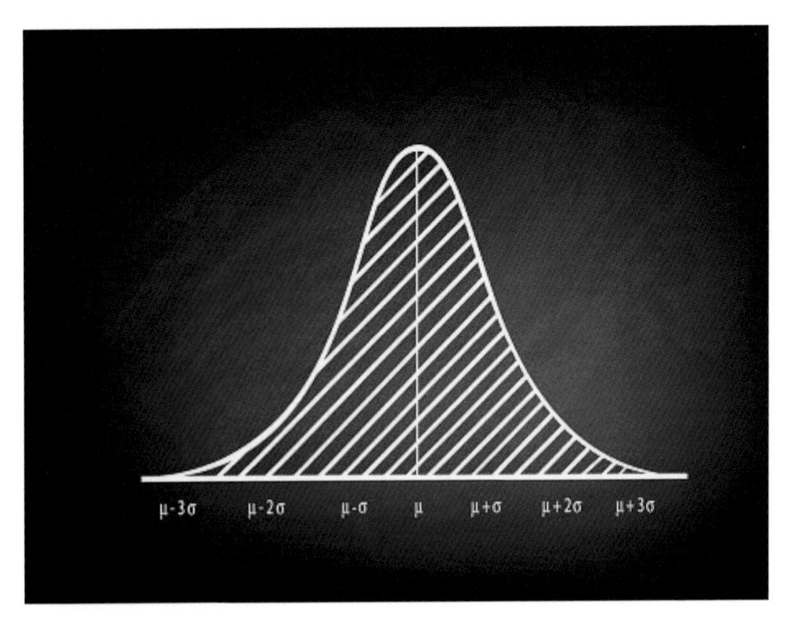

In a bell curve, the majority of the data is gathered in the center. The number of data points gets smaller the farther you get from the center. There may be some outlying data points that are on the extreme edges of the bell curve. This arrangement of the data is called its distribution.

Photo Credit © Getty Images/lamnee

Sir Ronald Fisher (1890–1962) has been described as the father of modern statistical science. Among other things, he is credited with the development of the t-distribution, which is used to analyze the distribution of probabilities (that is, where data is likely to fall on a bell curve) within a small population. You would use a t-distribution, for instance, if you were looking at the probable distribution of redheads in a group of thirty or fewer people.

Photo Credit © The University of Adelaide, via Wikimedia Commons

Nate Silver (1978–) is one of the best-known statisticians today because he correctly predicted the outcome of the 2012 US presidential election in forty-nine of the fifty states. His website FiveThirtyEight (referring to the total number of electoral votes cast in the election) is looked on as one of the most authoritative places for statistics about political races.

Photo Credit © randy stewart from Seattle, WA, USA, via Wikimedia Commons

The world is passing into an age of "big data," when enormous amounts of information will require statistical analysis by supercomputers such as the one seen here. Such computers are capable of reviewing trillions of data points in an extremely short time. However, the continued growth of data will require such computers to become even more sophisticated in the future.

Photo Credit © Getty Images/baranozdemir

key ones to look at: others are used in casual studies such as news reports, but they aren't considered to be scientific in nature.

What do 90 percent, 95 percent, and 99 percent confidence intervals mean? Keep in mind that two standard deviations of data under a bell curve represents 95 percent of the observations of that data. Ninety-five percent of the observed data will fall within two standard deviations, and 99 percent of the data will fall within three standard deviations under the bell curve. This is where the "5 percent margin of error" or the "1 percent margin of error" comes into play. (This margin of error is easy to find using statistics software, such as R, Python, or Microsoft Excel.)

If you've collected the data and run the numbers in statistical software, and the confidence interval comes back as 5 percent, then you know 95 percent of the time you can be assured that your sample is within 5 percent of the true proportion in the full population of the study. Remember, the newspapers are performing a study with a smaller sampling of the entire population. The entire populations of New York and Chicago are too large to test comprehensively, so a smaller sample of New Yorkers or Chicagoans was used for the study. Next, pollsters need to see if the quality of their sample corresponds to the makeup of the entire population. Finally, they will calculate the confidence interval. If it's within 95 percent, they're good: 95 percent of the sample set matches the larger full population.

MARGIN OF ERROR

When the newspapers are stating a margin of error, what they are reporting is the confidence interval in reverse. When they run the numbers, and the confidence interval comes back at 98 percent, then the newspaper will report a 2 percent margin of error. The next thing to ask is, "Is this confidence interval too far away from 100 percent and therefore significant?" Well, the answer is "It depends." If the poll is about vacation takers in Chicago, then a 5 percent margin of error is just fine.

MEASURING CONFIDENCE INTERVALS

Calculating Sample Sizes for Accuracy

When you are conducting a sample test of an entire population, you can either calculate the confidence interval of the test, or you can find how big your sample set needs to be to get to the level of confidence interval or margin of error you'd like for your study. Using statistical software, you can easily find the needed size of the study or the size of the sample can easily be found.

AN EXAMPLE OF MEASURING CONFIDENCE INTERVALS

It is good to know how accurate a sample set of a group is—especially if the sample set is a larger group of 100 or more observations. This type of question comes into play if you've administered a survey. Suppose you are trying to find out how many freshmen on a large university campus are 100 percent sure of their college majors. You know that many college freshmen come to school with a specific end goal in mind: a career, a degree, or enough credits to transfer to a more competitive school. You also know that on your campus there are 2,000 freshmen. You'd like to know what proportion of them are solid in the college majors that they've selected, but it would be too costly and time-consuming to ask every single one of them. Instead, you set up a random sampling of 100 students from each corner of the campus. You give them a survey and record the answers.

Your finding is that out of 100 students surveyed, 77 like and feel confident in their majors: that is to say, 77 percent of students polled answer positively to the question "Do you know and like your college major?" But the poll included only 100 students. Now you must ask yourself, "What is the quality of my survey?" or "What is the confidence interval of the survey?"

With statistics software such as Excel or R you can convert the 77 percent of 100 to *x* percent of 2,000. How is this done? The software will perform complex algebra internally to compute that with this survey, you could have a 90 percent confidence interval that the range between 70 percent and 84 percent of the population of the entire student body of freshmen would answer that they were decided and were satisfied with their majors.

Colleges commission studies like this all the time: they know that while signing up, the students act as if they know their majors in order to be able to get into classes next semester. But what the administration really wants to know is "How many of these students are happy with their majors/know their majors at this time?" That is, they want to know how many students are likely to switch majors.

The secret of knowing about confidence intervals is not just the math behind the numbers: even the simplest handheld scientific calculator can figure the number quickly and accurately. As with much of statistics, it's not crunching the numbers that tells you what you need to know. Rather, it is knowing how to interpret the information that is generated. Doing the math and finding the numbers is one thing. Getting to the point where you know what those numbers mean is totally different. Once you have the numbers, or the shape of a graph, it is up to you to know how to interpret them.

Once you've collected the information from your sample survey set, you'll need to ask yourself what confidence interval you'd like the range of the test to be set at. Ninety percent? Ninety-five? Ninety-nine? Of course, there are other levels of confidence intervals, but these three are favored by most studies.

GETTING TO 99 PERCENT ACCURACY

To get a higher level of confidence, you may need to change your sample population, your survey questions, or both. Suppose the college administration tells you it wants the test to have a 95 percent confidence level. Your previous sample size of 100 students gave you only a 90 percent confidence level.

After analysis, you determine you must have a sample of 367 to get to this level of accuracy. Try it for yourself!

Helpful Website

Here is a great site that has a few built-in statistical software packages in very easy-to-use formats: www.calculator.net/sample-size-calculator.html. Enter all the information you have—95 percent confidence level, 5 percent margin of error, population of 2,000 students, then let the software crunch the numbers.

Here's another one to try with the www.calculator.net/sample-size-calculator.html site. You've used the same survey, but your goal is results with a 10 percent margin of error, with a 90 percent confidence interval. The student population in this instance is 6,000. The answer shows a needed sample size that is much smaller than the 367 for the 95 percent confidence interval for the larger freshman class.

THE BASICS OF HYPOTHESIS TESTING

Positive and Negative Proofs

To decide what type of statistical data you need to collect, you'll need to circle back to the question you're studying. This question, what you are trying to prove or disprove, is called your *hypothesis*.

This hypothesis needs to be tested. A hypothesis comes in two forms: positive and negative.

The first is called the *null hypothesis*. The null hypothesis is a statement about the entire test population and about a quality of the population. If a statistical analysis is performed, and the data contradict this statement, then the statement is deemed false and is disproven. If the statement cannot be proven false then we fail to reject the null hypothesis.

This procedure is exactly like our legal system. The guiding premise is that the defendant is innocent until proven guilty beyond a reasonable doubt. If reasonable doubt is not shown, the defendant is deemed not guilty. The only time a defendant is deemed innocent is when the true perpetrator of the crime is found (and since this is not a television show, it rarely happens).

In the case of statistics, the null hypothesis is related to the quality of the test population because if you can't use the data to accept the null hypothesis, then the data might not have been meaningful in the first place. All you are saying is that the data either is enough to reject or not. If you fail to reject the null hypothesis, you are in effect saying the data was good enough to get you to that point. At the same time, you can't definitely say the other about a rejection of the null hypothesis. You can only say that this data isn't good enough—many people continue testing or redesign the test at this point.

It is best to keep in mind that the reverse of logic is being used here. Never in a study is it assumed that you accept the null hypothesis. Quite the opposite is true: you do not reject the null hypothesis. This is a subtle difference but an important one.

Null and Alternative Hypotheses

Keep in mind that the null hypothesis is a claim that you are testing. The alternative hypothesis is the complete opposite of the null hypothesis.

To accept the null hypothesis you are saying you are 100 percent certain that your predetermined ideas of how the study will turn out are true. You've asked a question, created a probable answer about how the question will be answered, set up the data, and then done the statistical tests. After all this, you need to 100 percent totally prove that the answer you discovered (the one you guessed in the first place) is true. You've accepted the null hypothesis. On the other hand, if you run all the tests but have yet to find that your hypothesis is not right, then you would not reject the null hypothesis. You simply don't have any evidence to say it isn't true. Therefore, you would be safe to state that the null hypothesis was accepted. Why is that important? Why is the wording of this acceptance/rejection important?

WHY USE REVERSE LOGIC?

With statistical testing, you collect data from smaller sample groups. If you were to measure entire populations, you would get 100 percent accurate results of the number of observations. This is often not possible, so you take a smaller sample set. When you administer a survey (or any other test) to this smaller sample set, you measure its level of significance. Look back at the graph of the confidence interval. The area outside that region is the level of significance. A bit about notation: the region designated as the level of significance is referred to as α, while the level of confidence is $1 - \alpha$.

One Tail or Two?

The null hypothesis can be one-tailed or two-tailed. A right-tailed test is designed for only positive acceptance, and a left-tailed test is designed for only negative acceptance. A two-tailed test is designed for both positive and negative outcomes. One-tailed tests are used in studies when a precise idea of rejection is important to only one end of the bell curve; for example, new medicine test rejection rates. In these cases, extreme acceptance rates to the medicine aren't as important as extreme rejection rates.

Five Steps to Hypothesis Testing

1. Find the question you want to study.
2. State the assumed answer to the question (called the null hypothesis).
3. State what the answer to the question is if the null hypothesis is proved wrong (alternative hypothesis).
4. Find the statistical test that you will use.
5. Determine the rules that would make the null hypothesis rejected or called wrong.

DECISION RESULTS

Let's continue to look at the legal system to help with our understanding of hypothesis testing. We have a defendant accused of a crime. Our legal system operates under the premise that the defendant is innocent until proven guilty. Consequently, our null hypothesis is that the defendant is innocent, which makes our alternative hypothesis that the defendant is guilty. The trial proceeds, all witnesses are sworn in and give their testimony, all evidence and closing arguments are presented, and the judge sends the jury to deliberate. Excluding the notion of a hung jury, there are four possible outcomes to this trial:

1. The defendant is innocent, and the jury delivers a verdict of not guilty.
2. The defendant is innocent, and the jury delivers a verdict of guilty.
3. The defendant is guilty, and the jury delivers a verdict of not guilty.
4. The defendant is guilty, and the jury delivers a verdict of guilty.

Clearly, in scenarios 1 and 4, the jury made the right decision, but in the other two cases the jury made a mistake. Here is the big question: which of the two incorrect decisions is worse? Scenario 2 violates the innocent until proven guilty tenet of our legal system.

From the world of hypothesis testing, this is the error we try to guard against, and we call the probability of making this error the level of significance. Depending on the nature of the hypothesis being tested, working with a level of significance of 5 percent or so could be fine, but there are issues, especially medical issues, where the level of significance must be much smaller.

TAKING IT TO THE NEXT LEVEL

The Quality of Large Sample Populations

If you are testing a sample of a set because the population is too large, it is often good to know if the values of that sample portion equal a constant or a repeatable number in the test.

In the test we talked about earlier that involved learning whether freshmen college students were happy with their majors, the experiment was based on a random sampling of the students. This was done by asking a sample of 367 students from all ends of the campus. To get more random results, the surveys were taken at three different times: 8 a.m., noon, and 2 p.m. In theory, if the test were repeated, it would show roughly the same results. In this experiment, the answer is binomial; the interviewer asks college freshmen, "Are you happy with your major so far this year?" The college freshmen can answer the question with only a yes or a no.

In this test you are testing for a specific confidence level and a specific margin of error.

Software

Statistical software can take much of the math burden off the user. In the old days, the math behind statistics was an arcane-seeming series of Greek letters and algebra, square root functions, etc. Nowadays, all you need to do is use the input boxes, and the software will crunch the numbers for you. Getting the right software is easy. Knowing the theory behind it is the hard part.

One of the bits of information you would need is which parameters you have determined in order to reject the null hypothesis. Remember, before taking the poll, you need to come up with a hypothesis. This hypothesis is your best guess of what you think the answer will be, and this is what you are attempting to not reject. By conducting the polls and analyzing the data, you know the percentage of your sample set that said yes. If you've gone to the software (such as www.calculator.net/sample-size-calculator.html) and you've found out how large your sample size should be to get your test to your required margin of error, and you've tested that many students, you can compare your average to your null hypothesis.

Try to break it down: the question is "Are the freshmen happy with their majors?" You assume the answer will be that 80 percent of freshmen are happy with their majors. This is your null hypothesis. Take the following steps:

1. Decide what level of confidence or what margin of error your test needs to be.
2. Use the software to find out how large your sample set needs to be.
3. Design a very random, repeatable testing of the entire population—but only the sample set.
4. Match your percentage of yes answers to the null hypothesis.
5. If the percentage of the yes answers is higher than your null hypothesis, do not reject the null hypothesis.

After you've run the test at the 90 percent confidence interval, and you are testing with a 10 percent margin of error, you will either end the test by rejecting the null hypothesis (which would mean the average number of students happy with their majors is below your original guess), or you would not reject the null hypothesis (which would mean your test sample came in with a higher percentage of yes answers than your original estimate).

You then have a choice: you can either stop the experiment there, call it good, and write up your findings, or you can take the test to the next level of confidence level. Why would you do this? In some scientific studies, taking it to the next level shows a completeness and an attention to detail that will be looked on favorably by those who read it. Some statistical studies can be done at the 90 percent level, while others must be done at the 95 percent or even the 99 percent levels or higher. In the case of medicine and testing new medicines, the statistical part of the studies is done at quite high levels, since it deals with health and well-being.

Taking it to the next level means going back to the software to find how large your next sample set needs to be. Use the software, find the number of people you need to test, and use the exact same testing procedures. Test a larger pool with the same questions, manner, time of day, and locations on campus: you are trying to repeat the experiment with a higher level of precision, not change it.

MEASURING LARGE SAMPLE POPULATION PROPORTIONS

When Data Gathering Runs Haywire

Suppose you run the experiment referred to in the previous section, the one about the percentage of students at a college who are happy with their majors. You run the experiment with a binomial structure at a 90 percent confidence level and then at a 95 percent confidence level. You change the number of students you ask, but you *don't* alter the questions, times the survey is administered, or the locations where you survey. You're just polling different-sized sample sets of students. What would you assume if the two polls showed a huge difference in their results? Even if both tests resulted in answers of yes at a high enough level for you to not reject the null hypothesis, but the percentages were vastly different, what would this mean?

The differences could be because both experiments were done with a random sample set; this means there's a degree of randomness through the experiment. Your different results may be the result of chance.

If that's the case, it means you could easily get an answer of 60 percent one time and 70 percent the next time. If your margin of error was 10 percent, you're still within it—that is to say, the percentage difference between the two tests is 10 percent or less. Remember, confidence interval is the reverse of margin of error. You can say that this is the possible swing in answers that could come about when you move from one size sample to another.

THE AVERAGES OF THE TESTS

There may be times in a statistical test that the variance you get between samples is greater even than the margin of error. Rather than 10 percent, the numbers differ by, say, 20 percent.

What would you do to correct your experiment?

First of all, look for an outlier, which may be throwing off the percentages. Remember, an *outlier* is an event that doesn't happen very often and is far away from the mean. If you see such an outlier, you can discard it and recalculate the variance.

If you can't find something that's causing the problem, you are by no means finished with your experiment. You'd need to repeat it with exacting standards: same location, same time of day, even with the same pollster, if need be. Remember, a key element to statistical studies is that they need to be repeatable. It is very common for contemporaries to rerun experiments you've done, first to see if they can be replicated, and second to see if you are reliable in your write-ups. Once you gain a reputation for your tests being repeatable, you are well on your way to being known as a reliable statistical tester, one who performs well-thought-out experiments.

The Importance of a Good Write-Up

The write-up of all the procedures that you've taken in the setup of the experiment—including the repeating of the polls, the statistical method of smoothing the data, and your conclusion—is one of the most critical areas of the presentation of your results. This is the section that your peers will look at the hardest and try to debunk the most. This is the section of your write-up that needs to be of the highest quality and the highest ethical standard.

If your data has gone haywire, and you've found that at the second test your numbers are way off, you'll need to repeat the test multiple times and then take an average. Just like the bell curve has a center around the mean, you can take an average of your tests to smooth out any outliers. Once you've taken a series of repeated tests, you can rule out any test that has results that are far from the average. You can then take only the center of the averages to eliminate the outliers entirely. In effect, you've taken a sample of samples.

THE HYPOTHESIS TEST

How Are the Decisions Made?

We know the four possible outcomes for making the decision about the validity of the hypothesis—that is, whether we reject or do not reject the null hypothesis. But we still have not addressed the criteria under which the decision is made. There are ways in which this can be done: comparing the sample data results with the hypothesized values or comparing probabilities. Let's look at the sample data process first.

CRITICAL POINTS

The key concept in the creation of a confidence interval or in the test of a hypothesis is the distribution. We never expect that our sample statistic will match exactly with the population parameter and accept that there will be a difference, within reason. *Within reason* is a subjective concept, but the usual application is that the results are within a certain range of the tested parameter.

Six Sigma

Six Sigma is a data-driven process that strives to eliminate defects in production and processes. The goal is that all manufacturing outcomes will lie within three standard deviations of the specified mean. CEO Jack Welch (1935–) applied this concept to General Electric in 1995. The name "Six Sigma" comes from statistical terminology used to study manufacturing processes.

How does this work?

1. Someone makes a statistical claim, e.g., the mean thickness of the washers we produce is 0.2 cm.
2. A random sample of washers is collected, and the mean thickness of the washers is computed. For the sake of argument, let's say the results of the sample of 500 washers are that the mean thickness of the sampled washers is 0.23 cm with a standard deviation of 0.04 cm.
3. "Is there something wrong with our production process?" the quality control manager asks. "Bring in the statisticians."
4. The null hypothesis is that the mean thickness is 0.2 cm, and the alternative hypothesis is that the mean thickness is actually larger.

The criterion is established that the company wants to be 95 percent sure that it is right. In other words, they are willing to accept a 5 percent chance they are wrong (the level of significance). They must now find the point that defines the maximum reasonable deviation from the mean. To do so, they look at the normal distribution with mean of 0.2 cm and a standard deviation of 0.04 cm. (Why the sample standard deviation? Because it is the only measure we currently have.) They determine the point at which 5 percent of the area lies to the right.

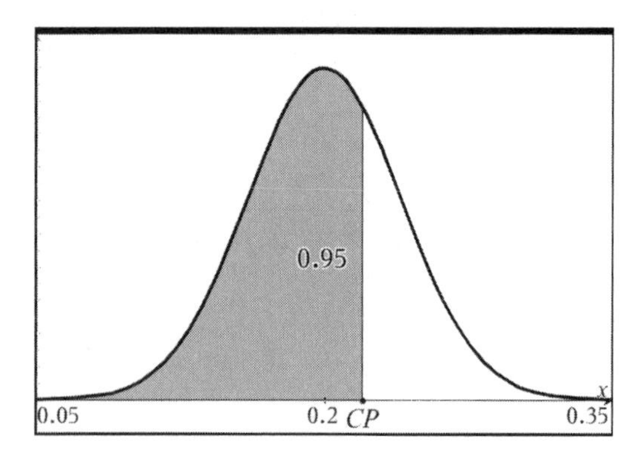

Technology is used to determine that the value of the critical point is 0.203.

Is the sample mean significantly large enough that the manufacturing process is deemed to be functioning improperly? Yes. The sample mean is outside the interval that has been determined as an acceptable deviation from the projected mean.

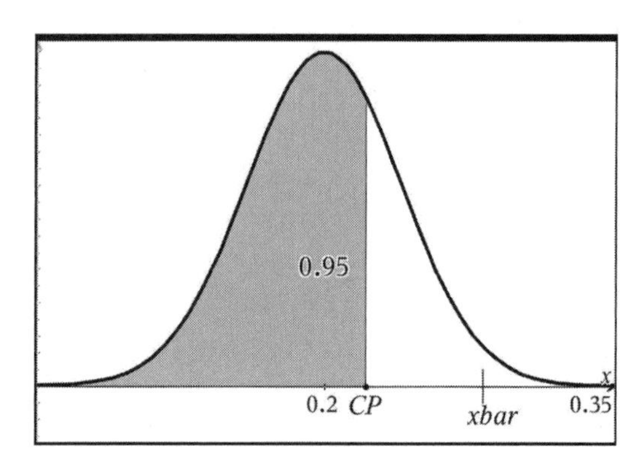

If the sample mean had been 2.01 cm, the graph would have looked like:

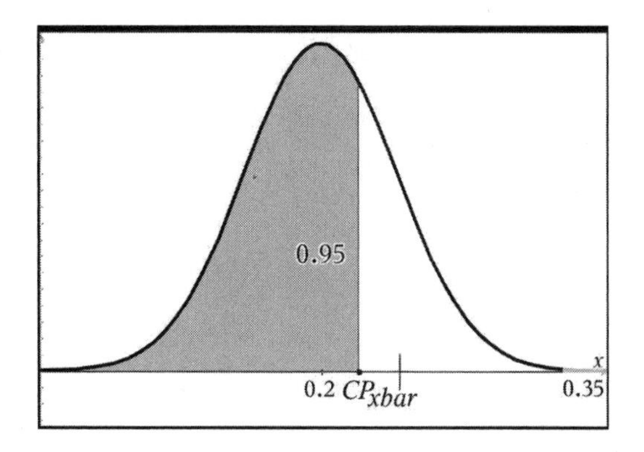

and the proper decision would be to claim the process is malfunctioning.

P-VALUES

With any study you are examining a null hypothesis and determining whether it should be rejected or not rejected. The P-distribution, or the P-value, is another method that can be used to help you determine the quality of your testing. The problem with the P-value is that the math behind the calculation is very complex and can best be done only with statistical software. This P-value is a measure of how small the significance level needs to be to make you reject the null hypothesis; that is, the P-value is the probability of an observed (or larger) result assuming the null hypothesis is true.

If you do not reject the null hypothesis, it is considered good at that point—until evidence comes along that shows it isn't good. With the P-value test, the statistical software through which you are running your tests will help you find what the smallest confidence level is to force you to reject the null hypothesis.

The P-value calculates this number the other way around. The P-value of the test will tell you at which confidence interval the null hypothesis must be rejected. In other words, your software might calculate the P-value to be at the 2 percent level of significance before the null hypothesis will be rejected.

OTHER MEANINGS OF THE P-VALUE

The actual number of the P-value has a strong correlation to the quality of the tests. P-values usually come in set ranges—this table will show how you can interpret the numbers in the P-values.

P-VALUES AND THEIR MEANINGS	
REPORTED P-VALUE	**MEANING OF THE NUMBERS**
P-value > 0.1	P-value points to nothing against the null hypothesis
0.05 < P-value ≤ 0.1	P-value has some evidence to reject the null hypothesis
0.01 < P-value ≤ 0.05	P-value has good evidence to reject the null hypothesis
0.001 < P-value ≤ 0.01	P-value has excellent evidence to reject the null hypothesis
P-value < 0.001	P-value provides the highest evidence to reject the null hypothesis

But in this case, you are telling the software what the level of significance is, and the software is *telling you the level of significance needed to be at these levels of rejection.* Using this chart, you can look at the reported P-values and determine if you are able to not reject or to reject the null hypothesis. Thus, you can see that the P-value is a very handy way to measure the quality of your tests and the quality of the data that you've collected. The P-value is telling you the confidence levels you need to be at to get to the point where you can reject the null hypothesis. It's like flipping the numbers around. You are asking the software to tell you what you need to get to before you can reject.

For example, let's go back to the case of the thickness of the washers. With the sample mean equal to 0.23 cm, the P-value is 21.3×10^{-63}. This tells us that there is almost no chance that the sample mean can be larger than 0.23 cm if the true mean is 0.2 cm.

Using the P-value won't change how you set up the study; everything will be the same. There is the same collection of data and the required number of data points. You'll still examine the data to see that the bulk of it gathers around the mean, though there will be a few outliers. Finally, keep in mind that if you perform a test and the numbers seem to be way off, and you've taken it to a smaller level of confidence interval (80–90 percent), then you'll need to retest a larger group. This would mean retesting another sample of washers. When you've done this, you'll measure the average, or the mean, using the P-test, and then you would decide to reject or not reject the null hypothesis.

Finding the P-values with statistical software can be an easy-to-use method of finding out if your tests can lead you to a conclusion in a quick, clean, and noncomplex manner. P-values are an accepted form of evidence in a statistical write-up, so they should be included in your executive summary, as well as the section in which you write about the procedures you used to test your data.

PATTERNS IN DATA

Bunches, Straight Lines, and Others

Once you have gathered all your data, you can then plot the up/down and across data on a chart. After you've done this, often you'll notice a telltale association within that data, how some of it bunches together. This section will show you how to recognize these associations and how you can interpret what they mean.

PLOT THE DATA ON A GRAPH

Once you've collected the data from your studies, if it is *bivariate* (meaning it has two variables, such as children's height/weight relationship studies) you can start to fill out a chart.

Bivariate Data

Bivariate data is just another way of saying the numbers have an *x*-axis and *y*-axis. Such data can be plotted on a chart that has one set of values (quantity, time, age, stock return, etc.) along the bottom of the chart and has the other set of values along the side of the chart.

Bivariate data is typical if you've collected data with which you're trying to find the relationship between two sets of information. For instance, let's say you took a poll, the object of which is to find out how many people take public transportation to work when there is a snowfall. On one side of the chart, you plot how many people took the train and bus, while at the bottom of the chart you show how many inches of snow have fallen.

In this example, each measure of snowfall inches is one observation of data. You might decide to scale the survey by 0.25 inches of snow, going from 0 inches to 16 inches of snow (a blizzard in some cities!). With this information, you plot the inches of snow on the bottom and match it up with the number of people on trains and buses on the side. After all the information is collected, you would probably notice that the data points (or each separate observation from each level of snowfall) form a line generally sloping upward.

What would this tell you? Well, from one or two or even three snowfall/public transport measurements, you couldn't really tell a whole lot. But after gathering many snowfall/public transportation observations and graphing them, you would be able to see a clear, ever-increasing amount of people using public transportation as the snow gets deeper. You might find that fewer people take public transportation to work when there is less snow, with the number of people increasing steadily as the snowfall gets bigger. (It's possible, of course, that in this graph you'd find a certain point where there'd be a sharp drop-off of people using public transportation, since if the snow gets too deep, people would not go to work and would stay home, out of the bad weather.)

How is this of practical value? Well, this method is exactly what major cities use during a storm to determine how many cars to add to trains, how many extra buses to run, how many extra drivers/conductors are needed to work: all depend upon the predicted snowfall in the city.

LOOK FOR ASSOCIATIONS IN THE DATA

In the snowfall/public transportation example, the evidence will probably indicate that as the snowfall gets deeper, more and more people take trains and buses to work. This is logical, because the trains and buses are safer and more reliable than driving on an expressway in a snowstorm. The fact that the deeper the snow, the more people take trains and buses is referred to as a *positive association*. More snow = more people on buses and trains. This positive association comes in two forms: perfectly positive association and very strong positive association. If the data were perfectly associated, then for each unit of snowfall, there would be an exact increase in riders on trains and buses. This exact increase would be the same for each unit of increased snowfall, moving together in lockstep. Since data in the real world usually doesn't happen this perfectly, there is the second association: very strong positive association. This manifests as a very tight, but not exact, correlation of the results.

Pearson Correlation Coefficient

The Pearson correlation coefficient, r, measures the strength of the relation between two variables. The range of r is between –1 and 1. A value of 0 indicates no correlation between the variables, while a negative value of r indicates that as one variable increases, the other decreases. A positive correlation coefficient indicates that as one variable increases, so does the other variable. As a rule of thumb, if $|r| > 0.7$, there is a strong correlation between the variables.

The opposite of this, of course, is a negative association—for example, if the data plot showed that fewer and fewer people ride the train and buses as the snow gets deeper and deeper. Just as with positive association, this comes in perfectly negative and very strong negative associations. Here's what these associations look like:

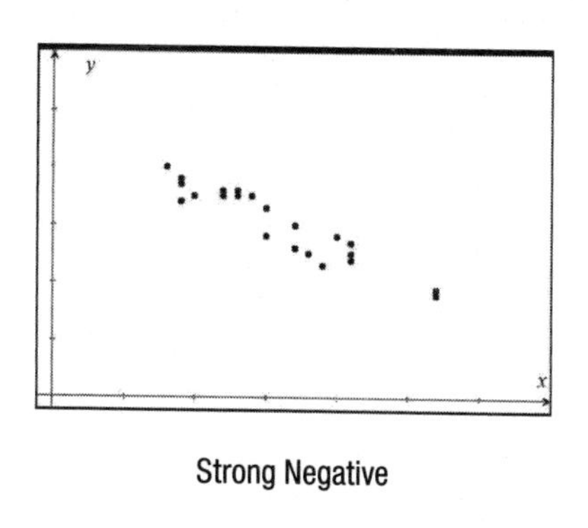

Strong Negative

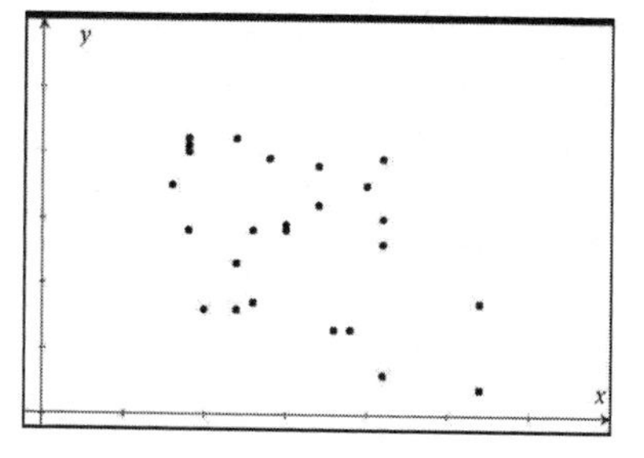

Weak Negative

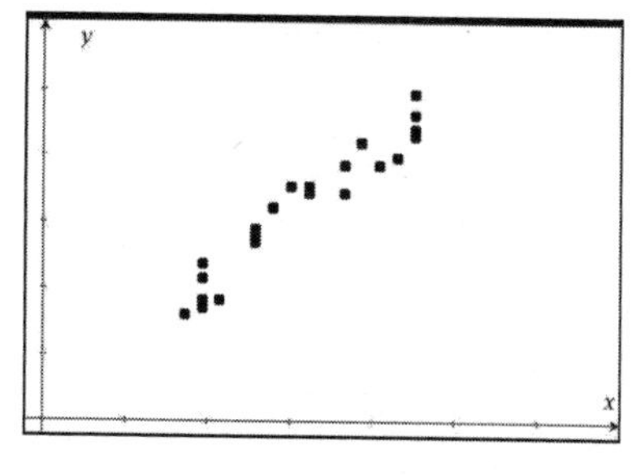

Strong Positive

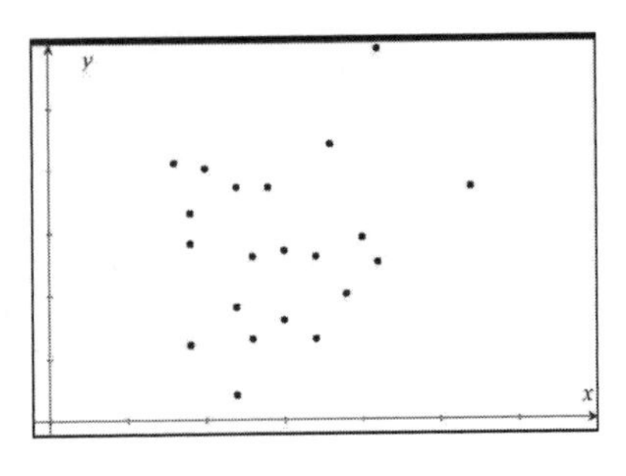

Weak Positive

The scatterplot of a relationship with strong correlation will be very tight, as the display shows. The more spread out the scatterplot, the weaker is the correlation. It is not reasonable to be able to determine a weak correlation from no correlation from a visual perspective—the correlation coefficient will need to be computed and a statistical test done on it to determine if there is a correlation between the variables. Finally, if you've done all your surveys and made many, many observations of train and bus ridership with many different levels of snowfalls; if you've plotted all the data on the graph; and if there is no obvious shape to the gathering of data, then you have *no association* in your study of snowfall/public transportation.

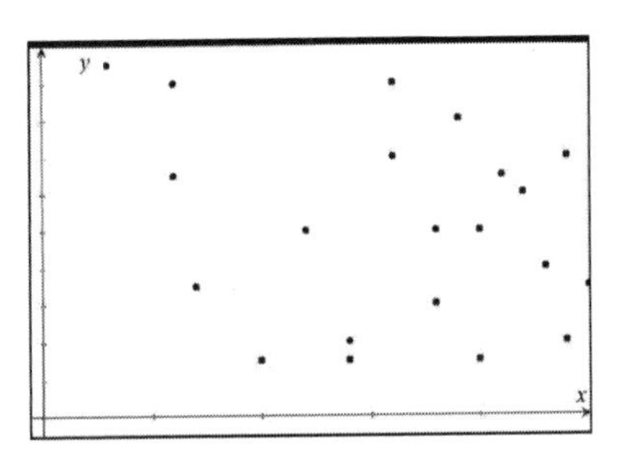

No Relationship

Why? Because the data plots are too random: the bus and train ridership doesn't seem to change in a predictable pattern with changing snowfalls.

SIMPLE LINEAR REGRESSION

If there appears to be a strong relationship between the two variables, you can attempt to determine if there is a predictable relationship between the variables using a process called *regression*. To keep the explanation simple, we'll take a look at a scenario where there appears to be a linear relationship between the variables (although we could also look at an exponential relationship, a power function relationship, or a logarithmic relationship in a similar manner).

A student of nutrition is looking at the size of the sandwiches served at a popular restaurant and the number of calories in each sandwich. The data collected is:

MENU ITEM	SERVING SIZE	CALORIES
⅓ lb Original Thickburger	343 g	770
⅓ lb Cheeseburger	240 g	620
⅓ lb Mushroom & Swiss Thickburger	259 g	650
⅓ lb Bacon Cheese Thickburger	320 g	850
⅓ lb Low-Carb Thickburger	245 g	420
⅔ lb Double Thickburger	445 g	1,150
⅔ lb Double Bacon Cheese Thickburger	436 g	1,200
⅔ lb Monster Thickburger	386 g	1,320
Six Dollar Thickburger	383 g	930
Little Thickburger	220 g	570
Little Thick Cheeseburger	167 g	450

MENU ITEM	SERVING SIZE	CALORIES
Charbroiled Chicken Club Sandwich	306 g	630
BBQ Chicken Sandwich	271 g	400
Low-Carb Charbroiled Chicken Club Sandwich	250 g	360
Big Chicken Fillet Sandwich	319 g	710
Spícy Chicken Sandwich	153 g	440
Regular Roast Beef	128 g	310
Big Roast Beef	171 g	400
Hot Ham 'N Cheese	131 g	280
Big Hot Ham 'N Cheese	232 g	460
Fish Supreme Sandwich	225 g	630
Jumbo Chili Dog	145 g	400
Double Cheeseburger	208 g	530
Small Cheeseburger	139 g	350
Small Hamburger	126 g	310

Source: "Fast Food." *Nutrition Sheet.* www.nutritionsheet.com/facts/restaurants/fast-food.

The student makes a scatterplot for the data:

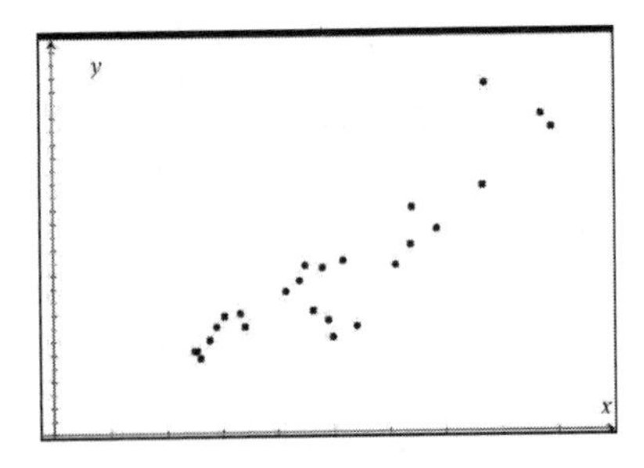

She uses technology with the data to determine the regression line that relates the size of the sandwich with the number of calories:

calories = $2.7 \times$ size $- 70$, with a correlation coefficient of 0.903. This is fairly strong evidence that the number of calories in a sandwich from the restaurant is tied to the size of the sandwich. (Not that anyone is really surprised by this, but the data seems to support our assumption.) Is this absolutely correct? There are statistical processes that you can perform, such as a hypothesis test, on the validity of the slope of the regression line to verify the findings. Once done, this equation can be used to predict the number of calories that will be in one of the restaurant's sandwiches. However, the size of the sandwich must be in the range of the data used. Since the smallest sandwich has 126 grams and the largest 445 grams, it would be fine to predict the number of calories in a sandwich with 325 grams but not in a sandwich with 610 grams. That is, we can use the regression equation to make a prediction with interpolation (within the range) of the data but not extrapolation (outside the rage).

PREDICTING THE FUTURE

Multiple Regression Analysis and Data Points

We use multiple regression analysis as the statistical tool to measure the cause-and-effect relationship among several variables and to see how they relate to a separate variable. If you have an experiment that measures how two or three observations affect a separate factor, you'll be able to use a little data crunching to find out how they interact with one another.

AN EDUCATION RESEARCH EXAMPLE

Let's imagine you're conducting a study about what types of high school test-taking environments most affect the overall test scores of students. You might measure such factors as room lighting, room temperature, how many hours before or after lunch the test was given, whether or not the test questions were reviewed in class the day before, and if the test was online (multiple question/true/false questions) or on paper (multiple question/true/false/essay questions). You know that some factors affect student performance more than others. With multiple regression analysis, you can determine through a mathematical relationship which one affects students the most. You might find that room temperature affects the scores somewhat, but negligibly, while the time of day of the test and whether the test was reviewed affect the students' test scores a great deal. In fact, you might find that a two-hour review made the average score go from C+ to B+, while the time of day the test was administered (say, after lunch when everyone was well fed and not hungry) added an average of 7 percentage points to the students' typical test score. With this information, you could review longer for every exam and only schedule them in the afternoons, as close to after lunch as possible: you would have mathematical, observable proof that you were ensuring the highest test scores for your students.

The Process

Regression is a process of taking each point of data and plotting it on a graph. The data points are spread over the graph, not in a random way, but with the up and across parts of the graph representing the relationship between the two data points. One of these is the time spent on review, the other is the scores on the tests.

For instance, let's say one data point is a 70 percent class average test score tied to a 20-minute in-class review period. The next data point, from a different class, might be an 82 percent class average test score tied to a 75-minute in-class review period, and so on. Once all the data points are recorded, you should be able to make an estimate of how sensitive the class average test score is to each additional minute of in-class review period. This relationship is what you discover by the regression analysis. This seems complex, but you can do it pretty easily with any good statistical software program. These programs plot the points and determine the level of sensitivity of each cause/effect you are measuring.

A WALL STREET EXAMPLE

Let's look at another example: a Wall Street trader is devising a program to automatically trade stocks in the US stock market. In order to do this, he has to accurately predict what might affect the up-and-down motion of the S&P 500, since this is an important stock market indicator. He might have an intuition that bonds; the European stock markets; the Asian stock markets; gold; the stock market's mood and temper; and foreign currencies such as the euro, the Japanese yen, and the British pound all affect the daily performance of the US stock market.

His goal is to find which of these parts of the world financial markets are most related to the up-and-down action of the US stock market. In order to do this, the researcher must find the past historical values of all of these world financial market components. By combing back through several years of databases, the researcher is able to obtain information on how each variable (bonds, currency, etc.) performed on each day they were traded. He then uploads that data into a software program. He makes sure that he has the data on at least 100 days of past trading, giving him a statistically valid sample.

Accurate Data Is Crucial to Building a Model

Because this trader is looking to map a financial plan, it is essential that he gets accurate data from a significant time frame and multiple sources.

Using the statistical method of multiple regression analysis, the researcher/trader can use this group of data sets (called the independent variables) and measure if and by how much the daily up-and-down change of these variables affects the US stock market's daily up-and-down change (called the *dependent variable*). The result will be a very powerful analytical tool that precisely measures how all those other world market factors affect the outcome of the US stock market.

How Is the Data Found?

The first step in the analysis is to find great data points. The trader could find his historical data easily on financial websites such as *Google Finance* or *Yahoo! Finance*. Using these sites, he can download the data directly onto a spreadsheet, including the dates of the trades as well as the values of the indicators. Of course, knowing which financial indicators to use is also key. (It might take some experimenting to get the best financial indicators that are robust enough to make hearty and accurate directional predictions possible.)

The same is true if you're conducting a marketing, medical, social media, or academic study—any study, in fact, that seeks an accurate prediction of how a group of factors affect another, single factor. With these studies, knowledge of what indicators to use as independent variables is a must, and sometimes a little bit of experimenting with the data sets is required to get a good, strong result.

THE *T*-DISTRIBUTION

Confidence Intervals and Tests for Single Population

In this section you will learn how to test confidence intervals for sample sets that are smaller than thirty samples. You'll also learn the standard deviation of small samples, which is called the t-*distribution*.

In most cases, the sample set you test will consist of more than 100 samples. In many cases, it may be much larger. Sometimes, though, it may be as small as 30. If the sample set is smaller than 30, then a different set of rules apply. Why? It is thought that 30 samples offers too few to get an accurate read, mainly because the group isn't large enough to have sufficient randomness, and the results may be biased.

The normal bell curve still applies, but in cases of sample sets being smaller than 30 you don't use the normal distribution. Instead, you use what is called the *t*-distribution.

The *t*-distribution has some characteristics of the normal distribution and some special ones.

Parts of the *t*-distribution are the same as the normal distribution:

- It has a bell curve shape.
- The bulk of the data is centered around the mean.
- Its median, average, and mode are all equal to zero.

However, the *t*-distribution differs from the normal distribution in certain important respects:

- The variance of the bell curve is always greater than 1.
- The *t*-distribution uses a concept called *degrees of freedom*.

DEGREES OF FREEDOM

The concept of degrees of freedom is similar to the concept of outliers. In a normal distribution curve, the bulk of the data is within one or two standard deviations, with up to 95 percent of the data being under that part of the curve. In normal distribution data sets, only 5 percent of the data is in the very far reaches of the curve. In other words, only 2.5 percent of the data will be at the extreme high and extreme low of the information gathered.

Student *t*-Test

William Gosset (1876–1937) was an employee of the Guinness brewing company. He developed the *t*-test as a method for assessing the quality of the Guinness product. When he was refused permission to publish his findings, he did so anyway under the pseudonym "Student."

With degrees of freedom, a certain number of data points can be outliers, too; this helps you decide the quality of the data and how accurate the bell curve is. However, these degrees of freedom can exist much more frequently than a normal outlier in a normal distribution. Degrees of freedom are used to help you determine how close to a normal bell curve you can get.

What's the difference? If your curve was standard, you would have three standard deviations. If, on the other hand, your bell curve is based on a small sample of 30 or fewer data points, and therefore you are using the *t*-distribution, you could calculate using four or five standard deviations. Because the sample is small, you can expect there to be more outliers.

BEWARE OF SMALL SAMPLES

All of this said, it makes sense that the larger your sample set, the more accurate your test will be. Remember also that experts recommend a sample set of at least 100, if not more, to ensure a nonbiased, more accurate read on the entire population that you are studying.

What happens when the largest sample set that you can obtain is smaller than 30? The first thing to do is evaluate your study. If you've established a hypothesis and designed an appropriate way to collect samples, in theory you shouldn't have any issues collecting enough data. But sometimes it happens that your study just doesn't give you enough data points: you've conducted your survey and you come back with fewer than 30 answers. What to do then?

One choice is to work with the data you have. If you are conducting a study for school or a work project, perhaps the margin of error generated by the small sample set can be overlooked. The other alternative is to use the *t*-distribution we discussed earlier in this section.

However, if you are conducting a test that can be modified, it may be possible to widen the parameters of your search. If your sample set is extremely small, then the test may have to expand to include a wider range of possible data hits.

For example, let's say you're trying to measure the stock returns on exchange-traded funds (ETFs) that are invested only in oil futures. You would like to prove that oil future ETFs move opposite to the direction of the stock market in general. Your boss wants you to do this study to see if he should be trading ETFs when the stock market sinks in value.

As you are collecting your data from your terminal, you discover there are only sixteen traded oil derivative ETFs that measure the price of crude oil. You could conduct a statistical study, but you know your boss's money is on the line. You know sixteen data points is a very small sample set of the entire worldwide oil market, and you don't want to risk getting bad data.

You decide to expand your search of ETFs. You then look at oil futures ETFs, as well as all the worldwide traded individual oil company ETFs—you go one step further and include the commonly traded oil company ETFs such as the ETF "XLE"—the value of which is based upon a basket of oil company stocks. Your expanded search of data now gives you a sample set of 107—enough for you to have a normal bell curve and therefore better data to use when building your statistical model.

GROUPS OF DATA

Independent Tests and Dependent Confidence Intervals

This section will introduce the concept of testing for a relationship between groups of data. The *t*-test, which we discussed earlier, is a test that can be used to measure descriptive statistics for like, yet separate, groups or descriptive statistics for same groups, both to measure differences.

This independent samples *t*-test could be used in the following example. Two teams from the same athletic conference are keeping track of how many touchdowns they make in each game.

TEAM NAME	GAMES PLAYED	GOALS
Eagles	25	70
Wildcats	20	74

The question that you'd like to solve is this: "Did each team perform relatively the same, or did one team perform better than the other and thus can lay claim to having a more prolific offense?" (You're not, of course, determining which is the overall better team; that would require a statistical analysis of the defensive performance as well as offense.) The null hypothesis is to assume that the average offensive performance for each game is the same.

Because you are using a margin of error of 5 percent, you are looking for events that happen only 5 percent of the time. If you look at a bell curve, 95 percent of the time the data will be covered, but what you would like to see are the two tails, each extreme end of the bell curve of the average or mean of the data (under the curve) that would happen 2.5 percent of the time at the smallest and 2.5 percent of the time at the largest ends of the bell curve (or the times the events were observed—in this case each time a game was played).

If an event came up in the 2.5 percent range on each side of the curve, you would consider it a rare event, and it would be beyond the 95 percent (and into the 5 percent margin of error).

Calculating and running the actual *t*-test involves very complex math and is best done with software.

Software's Triumph

In the past, complex statistical calculations were performed with pencil and paper. In a university class today, the algebraic functions of math calculations are still taught, but in the real world these calculations are done with software (like many of the calculations in this book). In fact, knowledge of the higher levels of math is falling to the wayside as a prerequisite to getting good jobs in data analysis and statistics. The higher-level math functions are being replaced with a deeper knowledge of how to use the software that, when programmed correctly, will crunch the numbers for you. However, you'll still need to know what a *t*-test is and how to use it.

After you've run the *t*-test, you can compare it to your margin of error numbers to decide if you can accept or reject the null hypothesis. In this case the null hypothesis is that the teams performed the same. In this example you conclude that if *t* was calculated to be smaller than –2 or greater than 2, you would automatically reject the null hypothesis. Running the *t*-test with software, you discover that *t* is 0.7. Because *t* is between the limits of –2 and 2, you accept the null hypothesis and state that yes, each team performed the same offensively per game, for the number of games each played.

DEPENDENT SAMPLES TESTING

If you have one group of test subjects, and you are testing them on if something affected them (for example, by how much it affected them), you can use dependent samples *t*-testing to measure the outcome of the tests. First, you'd come up with a null hypothesis. Let's consider the following example:

You're researching whether a school breakfast increases grade point average in grade school children. Your null hypothesis is that yes, the grade point average is not changed (the difference is 0), and the alternative hypothesis is the grade is increased (difference is greater than 0). You've tested ten children, and you've given them a simple math quiz. Using the same 5 percent margin of error, you use computer software to find that the range of the *t*-test must be between -2.25 and 2.25. If you use software to calculate the *t* of the group using the *t*-test and it is smaller than -2.25 or greater than 2.25, you can scientifically reject the null hypothesis and say that no, breakfast does not affect grade school children's scores on exams.

The first step is to administer the quiz to the ten students without the breakfast and record the grades. The second step is to give the students breakfast, and then after they've eaten, give them a quiz on the same subject and of the same difficulty and record these grades.

Subtracting the no-breakfast scores from the breakfast scores, you would come up with a new score per student, called the *difference score*. This difference score represents the change from one test to another (and this is what you'll measure to see if breakfast made a difference or not).

These new numbers are the ones that you'll use for your *t*-test. In this case, your software calculated *t* to be 3.6. Because *t* turned out to be out of the range of the *t*-test to accept the null hypothesis, and is at 3.6, well beyond 2.25, you can reject the null hypothesis and say that yes, breakfast did in fact affect the exam scores of school-aged children.

TESTS FOR TWO POPULATIONS

Small Sample Confidence Intervals

This section will explain how to test the confidence intervals for two separate population tests when both are less than thirty. As we've seen, when you're using small sample sizes, you can't use normal distribution, and therefore the normal standard deviations aren't used either. This section will help you learn how to estimate the standard deviations of smaller sample sets.

Sample sets that measure more than 100 pieces of data are considered normal and use the normal distribution curve. The normal distribution curve has the bulk of the data (95 percent) contained under two standard deviations under the bell curve. While this is very helpful to predict where the next data group or sample set would measure at, it doesn't help for small sample sets. Why is this true? Because with sample sets of fewer than thirty observations, you'll run the risk of not getting an accurate read of the entire population's true values. Remember, a sample set is a measure of the entire population: you are attempting to statistically analyze a very large group by testing a smaller group. You also know that the sample group must be random and large enough to be a good measure. A small set can be very misleading—there just isn't enough data to measure.

It can be very difficult to get a smaller sample set to center around the mean. If the data set is small enough, there may be no pattern to the data at all (making it useless). On the other hand, if a pattern forms in a typical bell shape, there may be so many outliers that the data is measured to more than three standard deviations. This is the problem: how do you measure the true standard deviations of such small samples?

SMOOTHING STANDARD DEVIATIONS IN SMALL SAMPLES

Keep in mind that when a normal, large-sized sample set is created, the goal is for the tests and results to be repeatable. The tests should be done using a random sample, but at the same time, they should follow the statistical rules so that the results you've drawn from the data can be repeated by others. Occasionally if someone performs a similar test to yours, it might result in a substantially different result, but most of the time the data will have very similar variances, and therefore the tester will come to the same statistical conclusions.

What happens when your data set or your sample set is so small that it is inherently possible that there will be many outliers?

You can cope with this situation by smoothing. In this case, if you've done a sample test and reached a conclusion, you come out with your version of the standard deviations, however many you need. Next, the other tester runs the same tests and comes up with her standard deviations. To smooth them, and get a more identical result, first take the square of your first test standard deviation, then take the square of the new test standard deviation, and add these two together. Calculate the square root of the sum of these two numbers. This has the effect of smoothing the data you and the other tester generated; the result is a more accurate average number that other researchers can use for comparison.

Pooled Variance for the Difference of Two Means

If s_1 is the standard deviation for the first sample with size n_1 and s_2 is the standard deviation for the second sample with size n_2, use the pooled variance $s_p^2 = \frac{(n_1 - 1)s_1^2 + (n_2 - 1)s_2^2}{(n_1 - 1) + (n_2 - 1)}$ with the t-test statistic $\frac{(\bar{x}_1 - \bar{x}_2) - (\mu_1 - \mu_2)}{\sqrt{\left(s_p^2 \frac{1}{n_1} + \frac{1}{n_2}\right)}}$.

The more samples you can take, the better the smoothing effect will work. Remember, small samples have multiple outliers, and this data smoothing technique can go a long way in helping you interpret the information. If you are taking the samples yourself, run multiple tests with multiple small batches. Each one is then squared and added together, and the square root is calculated.

Keep in mind that statistics is not an exact science. Its goal is to find meaning in data. The data must be of good quality, and sometimes this data needs to be "cleaned" in order to get more meaningful results. Giving data a good cleaning sometimes means removing the effects of occasionally weird outlier data. This can be done with averaging, but sometimes a more complex squaring/square root smoothing method works better.

STATISTICS IN ACADEMIC RESEARCH

Design the Quantity Around the Question

Statistical research never stands alone. Most statistical tests are done as part of larger overall studies. These can be part of business or political applications, but other times statistical studies are done for academic purposes.

For some bachelor's, master's, or doctorate degrees, a final project sums up all the work that you've done toward that degree. This paper is often called the dissertation, the thesis, or the senior project.

These papers often have a data component, referred to as the quantitative part. This term *quantitative* means the section contains numbers, and the conclusions of the paper depend on these numbers. This is the section of the senior project or the thesis that is often the most difficult for students to master.

If you're writing a thesis or dissertation, the first step you'll need to take in setting up the quantitative section of your paper is to decide on a question that you'd like to answer with the paper. Once you've done this, it is much, much easier to build a scientific gathering of data, design a statistical test, and then analyze the data to see if it fits your null hypothesis.

DESIGNING THE QUANT

The first place to start in any quantitative experiment is in the library. If you're not sure what to write on, but you know the basic direction you'd like to go in with the paper, then read up on the subject. If you read other academic papers, pay especially close attention to the summary at the beginning of the paper, as this section will tell you what question that paper is asking. This section will also explain a bit about the data collection methods, as well as the method of statistical analysis used and the conclusion.

If your paper is to replicate someone else's study, you can use her or his paper as a guide. If your study is on a new subject, other papers will help you see where researchers have gone before with their studies on that subject.

Once you've read up on the subject, your next step is to formulate your own question. This question will shape the entire direction of the quantitative study.

After you've asked the question, the next step is to come to an idea of how you think the question will be answered. This question will be answered with the null hypothesis. Collecting data is the next step. Designing a survey, collecting data off databases—either way, you'll need to have a good representation of the entire population that you are studying. If need be, you can look at a sample population.

DATA FOR A SAMPLE POPULATION

Sample populations can offer you a good cross section of how the entire population looks, but the key is to gather enough of the sample to ensure that you have a wide breadth and that the sample has been taken randomly. Random samples can be done in a variety of ways.

- **Secondary information:** This means you're collecting data from other studies or from other precollected databases. Since there are a wide number of professional, financial, educational, medical, and social databases available online, there's no lack of material for you to analyze.

- **Primary information:** If you are collecting data yourself, this is called primary data. Primary data is an absolute must with some subject matters, while with others secondary data is the normal type of data and is more acceptable. Either way, the data is collected with enough data points that you can run a statistical study.

Next, you'll need to decide on the level of accuracy, or the degree of error, you'd like to run the experiment with. Using software, you can input the size of the total population, then enter what degree of error you want, and the program will then tell you how large your sample set needs to be.

The next steps are to measure the sample or make observations and then run your statistical tests. From there you can determine if you are to reject or not reject the null hypothesis.

If the sample set looks to be way off from your null hypothesis number, you can rerun the sample set with the same degree of error and from there smooth the results. If your answer is to not reject the null hypothesis, you can either stop—you've had a good test and achieved your desired results—or you can rerun the test with a new randomly obtained sample set, but this time with a lower degree of error.

Either way, you stop, retest, and smooth or retest at a higher quality; then the statistical part of the test is done, and you're ready to write the part of the paper that describes all your steps and conclusions.

GETTING GOOD DATA

Sources, Quantity, and Quality Data Points

Conducting a multiple regression experiment is best done with quality information, and lots of it. The source of the information and what data is collected ties directly to the question you're studying and, more importantly, to your null hypothesis. With multiple regression analysis there are multiple null hypotheses—one each for each test of each independent variable. You're testing each one with multiple regression analysis, so the sources and the amount of data are key to reliable conclusions of the tests.

A multiple regression experiment is done to test the relationship between one fact and other facts, to determine how they affect it. Another way to look at it is to ask yourself the question "What elements can I observe that affect the outcome to this element? What can I add to or subtract from my study that I can also test to see how it affects the outcomes?" A second way to set up the question is this: "What factors of the experiment that are changing affect this single factor that might affect the outcome that I'd like to see also change?" A more scientific way of formulating this is: "What are the independent variables that might change (and by how much) the dependent variable?"

The scientific way of stating the question is easier to understand, especially if you know that the independent variables are the part of the experiment that are changing and you're measuring how much they affect the part you want to measure. This may seem like a roundabout way to explain multiple regression analysis, but these wordy, roundabout ways can help you visualize what is being done and what the question of the statistical experiment is trying to prove.

For example, you might have a good idea of what variable affects what, but you're not sure, so you'll need to run the statistical experiment a few times to see what is actually going on. Remember, a test that has no results doesn't mean you have to stop there: you can keep testing different things. Eventually you'll get to the point where you can say, "Now I see it. This affects that, and by this much." For example, a small business has a budget for advertising in newspapers, on the radio, on television, and online. Adjusting the amount paid to each medium will help in determining the best return (measured in sales) based on the distribution of advertising funds. So the business does a statistical analysis of how many new customers each kind of advertising brings in, and based on that study, it allocates its advertising dollars for the following year.

If you are looking to find out what independent variables affect the dependent variable, you'll need a deep, wide database to collect your information from. Knowing where to look is easy: here are links to some of the more well-known databases:

- https://fhssrsc.byu.edu/Pages/Data.aspx. This is a list of links to some public access databases that can be used for a variety of subjects.
- Financial databases include *BigCharts* (http://bigcharts.marketwatch.com/) and *Yahoo! Finance* (https://finance.yahoo.com/).
- Bloomberg also offers a free demo: www.bloomberg.com/professional/request-demo/. While temporary, it offers a great depth of information for a finance statistical study.

Download the data to Excel or other software.

SELECT YOUR INFORMATION

In every one of these databases, you'll need to navigate to the place containing the information you need for the test. If you're doing a multiple regression analysis, you can choose up to eight different factors to measure that could influence your target. What are the factors to measure? Again, this is where you'll have to come back to the question of the study and, more importantly, the null hypothesis. Remember the issue we discussed earlier concerning stock market returns? The null hypothesis is all the elements you think might affect stock prices: stock volatility, gold prices, currency prices, other world stock indices, etc. Remember, if you've crafted your question of study well, and you've crafted your hypothesis of the study, then you'll know exactly what data elements you are trying to measure.

YOU'VE FOUND YOUR DATA—NOW WHAT?

Once you've found your database, the next step is to find the information for each independent variable. Each variable you are testing needs its own data collection; if you are measuring five different elements to see how each one affects the target, then you'll need to perform and collect that information on each of the five elements.

For each element there should be a way to reference it—a date, reference number, or other relational way to tie that factor to the independent variable. In other words, you're attempting to measure how factor A affects factor B, so what was the relationship of factor A to factor B, measured in time, proximity, or some other way? This is a key element to the regression test. With our examination of stock prices, you look at the price each day. That's your structure for the analysis.

Finally, you import the data you've accumulated into Excel (or R, if you are using that statistical program). Your software can run the regression analysis and crunch the numbers.

The quality of the data depends on the database you are using. If you are using a reliable source, then your data will generally be reliable. Remember, trusting your source is key to building a test that can be replicated. Collecting data, running a regression test, and drawing conclusions is one of the key skill sets in statistics.

Multiple regression analysis is a powerful tool that can lead to the next step of statistics: predictive modeling. To build useable, reliable, and worthwhile predictive models, using this form of testing—a form of *back testing* (because you are using "back data," or historical data), you need reliable sources of information: governmental, commercial, or paid-for databases are the best. They are tried and true, exist to be used, and have the data cleaned often. Your test will have an extra layer of professionalism and acceptance if the data has been collected from a well-known source. The source is most likely used by other statisticians, and therefore the errors in the raw data have been worked out.

If you are building a test of multiple regression from a private database (from work or from your own studies), then the methods of random testing and large data sample sets come into play. Know your software and use it when it comes time to choose how large a sample data set to use—the larger the set, the better. Use smoothing techniques if the tests are from independent, private databases or if you've collected smaller amounts of data to test. Finally, know your null hypothesis, and don't be afraid to rerun the tests if none of the variables tell a story of affecting your target. If you've made a hypothesis, and you're now rejecting all or most of it, then rerun the regression analysis with new factors to test your conclusions. These tests, while time-consuming, are not difficult, because the software does the number crunching for you.

A REGRESSION EXAMPLE

Electricity Use in the Suburbs

Assume that you would like to know what factors affect the electricity usage in a suburban area. Your hypothesis is that such factors include rain and snowstorms (since people stay in their houses more); holidays (as people stay in their homes and cook more); and nights of music, movie, and TV award ceremonies (because people have watching parties at their homes and stay in).

Your first step is to determine the amount of electricity usage every day for the past ten years. This gives you the base, or the dependent variable, you are testing against. This data can be found on your electric company's website. You upload it into a spreadsheet with the dates attached. Next, you have to find the dates of each event that you are testing as an independent variable. You need to find the past ten years' dates of rain and snowstorms, possibly from Weather.com. You'll also need to find the holiday schedule for the past ten years, and you'll need to find when various award shows were broadcast on TV for the past ten years. You upload all this data into a statistics program, with each factor in a column on the spreadsheet, and each date's information on the same row on the same sheet.

I'll describe the next step as if you are going do the regression in Excel, but you'd take similar steps if you were using other statistical software such as R or Python. If you know the commands, it doesn't matter what software you are using: the results will be the same.

Next, highlight all the information on the spreadsheets and then go to the Excel function called "regression." Depending upon your version of Excel, you can look online for the exact place on the toolbar to find these commands (or if you need to, download an extra Excel tool pack, which is usually free). The regression tool on Excel will ask for the independent variables and the dependent variables. After entering this information, click the run button, and your computer will crunch the numbers. In this example, it will tell you the amount that each factor affects electricity use.

You might notice that while the weather report information is reported in inches of rain or snow, it is done every day regardless of the weather. This is good and leads to accurate information. On the other hand, the information about TV specials shows that there are specials once a month or so. This isn't enough information to get results. Neither is the holiday information. You'll have to go back and rethink your data gathering to reflect something that you can gather more data on. You might want to change the TV and the holiday schedule to one entry: days of the week. You can upload the days of the week to Excel and test for that.

How would this be done? Remember, Excel does regression tests for numerical and quantitative information. Sunday through Saturday is hardly quantitative, and one day's value isn't higher than the next. Because of this, what you could do is change your study even further and measure the electricity usage on weekends. You set up Excel to list the days of the week for the past ten years, and then tell the program to convert Mondays through Fridays to be 1 and the weekends (Saturday and Sunday) to be 0. With this function, you've converted nonnumerical data to a binomial function of a weekend or a weekday.

Once this is done, you can rerun the regression analysis with the new, modified data and the new null hypothesis. The software will tell you what factors affected the electric usage the most: the rainfall/snowfall and/or the weekends, and by how much usage was affected. The program will tell you with mathematical certainty by what factor of 1.0 the electricity usage was affected. If the experiment was run in Excel, and the regression showed that rain had a factor of 0.25 and day of the week had a factor of .72, then you would know that each inch of rain or snow increased the electricity demand by a factor of +25 percent, and each weekend day increased the electricity demand by a factor of +72 percent. (Other software programs would give you this information as well, but in somewhat different forms.)

Nate Silver

One statistician who rose to fame during the 2012 US presidential election is Nate Silver (1978–). He correctly predicted the election's outcome in forty-nine of the fifty US states, gaining a reputation for infallibility. This was shaken in 2016, when his website *FiveThirtyEight* gave Hillary Clinton a 71 percent chance of defeating Donald Trump in the race for president. Like most statisticians analyzing the race, Silver was wrong, but he continues to have a reputation for careful and usually correct statistical analysis.

From this the electric company can build a model: it can enter the expected rainfall, then add what day of the week it is, and come up with the expected amount of extra electricity demand for the day.

This is another example of how multiple regression statistics is done in real life. There may be other factors, but the inputs are constantly modified to get the study to the point where you can find data on it. You are building a model, and if your model is changed by changing the independent variables, that's fine. In this way you again are using the smoothing effect that is the norm in statistics. Remember: statistics isn't a pure science. It is the art of using math to interpret and predict outcomes.

WHAT REGRESSION DATA TABLES TELL US

Using Data to Build Predictive Models

Computer software will show you a lot of information from a typical multiple regression analysis, a method of finding the relationships between how events or observations affect one another. But keep in mind that while much information is given, only some of it may be required to get you to the point where you can accept or reject your null hypothesis. Even less of it may be necessary to build an efficient, simple predictive model.

Here is some of the data that comes from running a regression test.

R SQUARED

This tells you what percentages of the dependent variables that are centered around the mean are affected by the independent variables. If you got a 90 percent here, that means that 90 percent of the data is affected by the independent data.

R^2

The coefficient of determination, R^2, tells you what percentage of the output variable is explained by the input variable.

ADJUSTED *R* SQUARED

This number is used if there is more than one regression model. If you ran a multiple regression test, with multiple independent variables all at the same time, then this number would be your test number. It would act for each variable as the *R* squared would work for one variable.

STANDARD ERROR OF THE REGRESSION

This measurement tells how much error was allowed for in the overall test. It measures the precision of the regression. What you're testing for determines what level of error you're willing to accept. If you are to use your regression as a first step in building a model, then you will need to pay very close attention to this number. If you ran a regression of multiple variables and a few of them had a large error (meaning those variables tested with a very low precision level), then you would disregard these when building your model. For instance, if you're running a regression test in order to build a financial model—possibly the market level predictor test we looked at earlier—if you run ten tests and the standard of error on three of the ten is unacceptable, you should disregard these in building your predictive model.

We can't stress enough how important error levels and precision levels are. Your margin of error must be within acceptable limits if the predictive model you're building is going to work.

OBSERVATIONS

This gives the data of the raw number of observations in the test.

These second parts of the output from an Excel or R regression analysis aren't used that often, and you most likely wouldn't use them in model building, but they can be useful for a deeper and more complex statistical analysis.

SUM OF SQUARES (SS)

- *Regression mean square regression (MSR).* This gives the regression sum of squares/regression degrees of freedom, whereas regression Mean Squares is defined as regression mean square error.
- *Residual regression mean square error (MSE).* This gives the residual sum of squares/residual freedom.
- *Significance F.* This gives the significance level. Critical point for the F distribution is defined as $F = \frac{MSR}{MSE}$.

F Distribution

An F statistic is a value you get when you run an ANOVA test or a regression analysis to find out if the means between two populations are significantly different. It's similar to a *t* from a *t*-test. A *t*-test will tell you if a single variable is statistically significant, and an F test will tell you if a group of variables are jointly significant.

Here's how you determine whether to accept or reject the null hypothesis with your regression analysis.

Use the *t*-statistic to perform a two-tailed test of hypotheses to test the difference between the null hypothesis and the alternative hypothesis for each of the slopes in the multiple regression equation. You could also determine a confidence interval for estimating each of the regression equation's coefficients.

DETERMINING THE CAUSES

Getting the Details

We've talked about the steps required to run a multiple regression analysis. This section will help you further with the interpretation of the numbers that you will see when you have done such an analysis.

The first thing you'll need to do is perform the regression properly. Depending upon what piece of software you have, you'll need to find the regression command either in the data section or in another section of the software's tool bar, such as formulas. (You can use the software's help section to locate it.) Next, follow along with the prompts of the regression box, and once you run the regression an information page will appear in your worksheet. Of all the results that the software will produce, the main thing that you are looking for is how good the regression was. Does the information you were testing for make sense? Better yet, now that you've run the regression, can you use any of it to help you answer your question (or did it help you in your study question)?

This is key, because remember, you are using regression to measure a sample set of a population. And with any sample set, it is important to make sure the sample reflects the data as a whole.

To work the best, the sample set needs to be random. A regression analysis will tell you how good the quality of the sample set is, along with which of the factors in the sample set affect one another.

HOW TO INTERPRET THE NUMBERS

The first thing you'll want to look at is if the overall test results are based upon chance. In other words, when your computer software ran the regression, you'd like to see that you got results that were based upon quality, and that you're not just jumping to conclusions about the data. If the software says there is a high degree of chance, then in additional tests another researcher won't get the same results. So, the first thing you will look at is how much chance there is in the regression experiment. To do this, you will look at the *R* squared number that we mentioned in the previous section. The *R* squared number tells you the percent quality of the "guess" that the software made. Eighty-five percent or higher is a good score since it indicates that 85 percent of the output variable is explained by the input variable. If you have many independent variables (more than four), then you should look at the adjusted *R* squared numbers. Remember, we said this number is for multiple independent variables. It will give basically the same information as the *R* squared number, but the adjusted *R* squared is a more conservative estimate of quality of the regression and takes a larger regression analysis more into account. With an adjusted *R* squared a score of 80 percent or greater is considered good.

The next information to look at from the regression tables is the significance of F (also discussed in a previous section). This number will tell you each separate part of the regression's quality. Remember, the *R* squared and the adjusted *R* squared refer to the overall model, but the significance of F is for each part separately. In the case of this number, the smaller the better. Any significance of F that shows up smaller than 0.02 is good. There might be a variable that has a very large F number. You use only the smallest numbers in a model. Only the factors that the software tested under 0.02 should be used with a predictive model. If the significance of F in the analysis measures larger than 0.05, the data being analyzed should not be used in your predictive model, because the computer regression has shown they're not of quality.

The last thing you'll need to test is the data group. Go to the "Residual Output" section of the regression output screen. Highlight the "Residuals" section, go to the "Insert Graph" section of your software package, and use the residuals information that you highlighted to create a scatterplot chart from this data.

Once you create the scatterplot chart, you will look for two things: the data should be centered around the center (centered around the mean) and it should have a general, slight bell curve shape. If you find that's the case, the quality of the data is confirmed, and you can use it to build your predictive model.

CHI-SQUARE DISTRIBUTION

Measuring the Goodness of Fit

Chi-square distributions are used to test if a distribution has *goodness of fit*, that is, if the sample sets you're using closely resemble the entire population in their bell shape. It is often used for data that is not fully randomized and favors one end of the bell curve. The chi-square distribution can also be used to test the variance of a population.

Sometimes when a group of sample sets are taken and the data is plotted on a bell curve, the curve has a long tail to the right and a very short tail to the left. As we've discussed, the group of the sample sets under the bell curve is called the distribution. A *normal distribution* occurs when an equal amount of data points falls under each end of the bell curve and the bulk of the data is clustered around the middle. As we said earlier, normal bell curves also have one, two, and three standard deviations of data.

With the chi-square distribution, the distribution is skewed to the right for smaller numbers of degrees of freedom and becomes more symmetric with larger numbers of degrees of freedom.

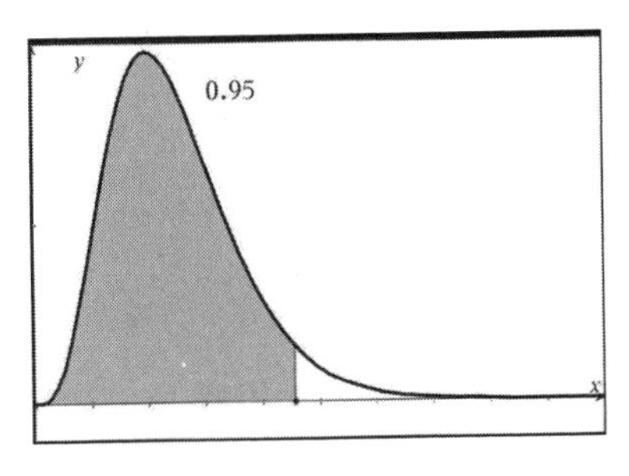

Image of bell curve with more data to the right

The term *goodness of fit* is used to compare the observed sample distribution of a categorical variable with the expected probability distribution. The chi-square goodness-of-fit test determines how well the theoretical distribution (such as normal, binomial, or Poisson) fits the empirical distribution. In the chi-square goodness-of-fit test, sample data is divided into intervals. Then the numbers of points that fall into the interval are compared, with the expected numbers of points in each interval.

The process assumes that in the null hypothesis there is no significant difference between the observed and the expected value. The alternative hypothesis assumes that there is a significant difference between the observed and the expected value.

Use the Chi-Square Goodness-of-Fit Test

The chi-square goodness-of-fit test is appropriate when the following conditions are met:

- The sampling method is simple random sampling.
- The variable under study is categorical.
- The expected value of the number of sample observations in each level of the variable is at least 5.

Based on this information, you can run your other tests: the mean, the mode, or even a multiple regression analysis. This is possible because you've used chi-square to measure the quality of the sample set. If the quality is good, then you know the sample set is a good representation of the entire population.

HOW QUALITY OF SAMPLE SET RELATES TO THE NULL HYPOTHESIS

If your initial test of the quality of the data is good, the next thing you can do is take a more scientific test of the quality to see if you can accept or reject your null hypothesis. You can use all of the other statistics tests to help you with the null hypothesis.

Chi-square distributions are a bit different than normal distributions. The distribution becomes more symmetric as the number of degrees of freedom increases.

In attempting to use the chi-square distribution to test the variance of a distribution, you need to know that the data is normally distributed. That should not be an issue because the central limit theorem assures us that the distribution of sample means will be normally, or at least approximately normally, distributed. The test statistic for the test of hypotheses is $X^2 = \frac{(n-1)s^2}{\sigma^2}$ where s^2 and σ^2 are the sample and population variances and $n - 1$ is the number of degrees of freedom from a sample of size n.

GOODNESS OF FIT

All tests of hypotheses consist of four steps:

1. State the hypotheses
2. Formulate an analysis plan
3. Analyze sample data
4. Interpret the results

We have a random sample of 500 US adults who are questioned regarding their political affiliation and opinion on how well Congress is performing. We will test if the political affiliation and their opinion on Congress are dependent at the 5 percent level of significance.

POLITICAL AFFILIATION AND OPINION ON CONGRESS PERFORMANCE				
POLITICAL AFFILIATION	**RESPONSE**			
	Favor	**Neutral**	**Opposed**	**Total**
Democrat	30	20	150	200
Republican	80	40	130	250
Independent	10	15	25	50
Total	120	75	305	500

State the Hypotheses

The null hypothesis is that the variables "political affiliation" and "opinion on how Congress is performing" are independent, while the alternative hypothesis is that there is a dependency between these variables.

EXPECTED COUNTS				
POLITICAL AFFILIATION	**RESPONSE**			
	Favor	**Neutral**	**Opposed**	**Total**
Democrat	$\frac{200 \times 120}{500} = 48$	$\frac{200 \times 75}{500} = 30$	$\frac{200 \times 305}{500} = 122$	200
Republican	$\frac{250 \times 120}{500} = 60$	$\frac{250 \times 75}{500} = 37.5$	$\frac{200 \times 305}{500} = 152.5$	250
Independent	$\frac{50 \times 120}{500} = 12$	$\frac{50 \times 75}{500} = 7.5$	$\frac{50 \times 305}{500} = 30.5$	50
Total	230	45	225	500

The degrees of freedom (df) is equal to (number of columns − 1) (number of rows − 1) = (3 − 1) × (3 − 1) = 4.

Test Statistic

The test statistic is a chi-square random variable (X^2) computed by finding the sum of the squared difference between the observed count and expected count. That is, $X^2 = \sum \frac{(O_i - E_i)^2}{E_i}$. In this case, the calculation will be:

first row: $\frac{(30-48)^2}{48} + \frac{(20-30)^2}{30} + \frac{(150-122)^2}{122} = 16.5609$

second row: $\frac{(70-60)^2}{60} + \frac{(37.5-30)^2}{37.5} + \frac{(152.5-150)^2}{152.5} = 3.2077$

third row: $\frac{(12-10)^2}{12} + \frac{(7.5-15)^2}{7.5} + \frac{(30.5-25)^2}{30.5} = 8.8251$

You can see why you'll let the technological tool do this work for you.

The value of $X^2 = 16.5609 + 3.2077 + 8.8251 = 28.5424$.

The critical point for a 5 percent level of significance and 4 degrees of freedom is 9.488. Our test statistic is greater than the critical point, so we have to reject the null hypothesis. The data supports the claim that there is a relationship between one's political affiliation and how one feels about how well Congress is performing.

TESTING VARIANCE

Another aspect of the chi-square distribution is that it allows for testing the variance of a normally distributed variable.

Here's an example of how the test is used:

A manufacturer believes that the diameters of the 2-cm ball bearings they make are normally distributed with a mean of 2.00 cm with a standard deviation of 0.005 cm. A sample of 25 bearings is taken, and the standard deviation is 0.007 cm. Is this an indication that the production process is not working properly? Use a 5 percent level of significance.

The null hypothesis for this test is that the standard deviation is less than or equal to 0.005 cm (we include "less than" as part of the hypothesis because the manufacturer will be comfortable with a smaller spread of deviations from the mean but not a larger spread). The alternative hypothesis is that the standard deviation is greater than the stated value of 0.005 cm.

The statistic for this test is $X^2 = \dfrac{24(0.007)^2}{0.005^2} = 47.04$. The critical value for a 5 percent level of significance and 24 degrees of freedom (found from a table of values in a statistics book or with technology) is 36.415. The manufacturer has to look at the production process because it is not working properly.

ANOVA BASICS

Analysis of Variance Tests

It is frequently the case that one might be interested in comparing differences in results among several groups. For example, a tire manufacturer would like to know how its best tire handles on different types of vehicles. To do so, the manufacturer would perform an analysis of variance (ANOVA) on the data collected. ANOVA is a method of analyzing four or more separate chi-square tests. If these four or other small sample sets have related data—for example, if they are all from the same questionnaire, from the same study, or answer the same questions in the same test—then the data is related. The data of the tests might be so related that you would say they overlap. This might be because several small group samples were taken with the same end in mind: to measure the same variables.

This would come into play if you were running an experiment with similar yet different target groups. In this way, you would run the chi-square tests on each group each time, and then run an ANOVA test to further smooth the data that you received from all four or five chi-square results. This ANOVA testing acts as a data smoothing technique, as it is intended to give reports as to the averages of means of four or five separate tests.

Keep in mind that statistics is not a hard science, and that there is a lot of leeway to use smoothing techniques when data is sparse. Also, keep in mind that with smaller groups of test data, or small sample sets, the data from each test will be very irregular. This is because the test data has numbers as small as 40–50 or even smaller. You've learned that sample set testing works best when the sample set has 100 data points or more. A larger size leads to more random results, which also leads to fewer irregular observations.

If you've run a small sample set test in your study, and although the numbers look good you'd like to reinforce your conclusions with further tests, then use the ANOVA tests on the data.

ANOVA versus *t*-Test

You use a *t*-test to compare the means of two populations and ANOVA to compare the means of multiple populations.

Running an ANOVA test is simple on most statistical software. It's a great test to use if you're stuck with smaller sample sets, and you'd like to run those tests multiple times. You do this to test any imperfections in the data that are too irregular. A good way to use this is to run the small sample set four or five times, each one separately with chi-square, and then use the overlapping technique of the ANOVA testing. This has the effect of finding averages of means, softening the effects of outliers, etc., all done through a layering effect.

When deciding upon which tests to use, it's best to ask yourself what the end goal is with your test. You'll learn very little by running ANOVA tests with no data. The best way to learn is to have a use for it: if you need to strengthen your results before you draw a conclusion on your data, then use it. If you need to gather information, test it statistically, and then use the results to either accept or reject your null hypothesis, then this test can help you be more confident in your results.

Remember, any statistical test uses data and methods that can be repeated by other parties. You don't want to make a conclusion about a population by running a small sample set, only to find that when others make the same tests, they come up with entirely different results. There is an art to analyzing data and drawing conclusions, but this art must be backed by sound science as much as possible. In this way, consider using ANOVA testing when you run into smaller sample set testing. It will strengthen your results, and you will feel more confident in your conclusions.

ANOVA AT WORK

One-Way Analysis of Variance

One of the more common tests of ANOVA is the testing of a hypothesis about the mean of a group of tests. ANOVA can be used to test a question such as "Are the averages of these groups the same?" This section will show an example of ANOVA at work, as well as comment on upper-level, complex calculations in today's modern statistics.

Let's use a rainfall example for our test. In this example, a large commercial corn and soybean farm of 200,000 acres would like to conduct a survey of average rainfalls during each month. They'd like to do the research going back tens of years, but for practical purposes they'll measure the rainfall in this year alone, for each growing month, and for each of the counties of the states that their farms operate in. Since there are hundreds of acres of land, they will select only thirty days out of each month this year to measure, and they have only twenty counties to measure, so it can be said that they are looking at a small sample set. How so? Measuring thirty days per month for twenty sites constitutes a small sample.

This test is called a one-sided ANOVA test because they're testing only one thing: either the county had rain, or it didn't; there can't be negative rainfall. Also, they have 200,000 acres in the population, but they'll be measuring the rainfall at only twenty sites.

The company would like to know how much fertilizer they need to order, so they'll need to know the average rainfall over the year (to order the fertilizer for next year).

They collect the data: twenty sets of data with only thirty days of possible rainfall out of each month.

Using one-way ANOVA, they can use its smoothing effect to find out if the averages for each of the twenty sites where they are measuring rainfall are the same. In other words, they have an average rainfall for each of the twenty sites: they've measured it throughout the season, and they think the average rainfall is four inches of rain—this is their hypothesis. ANOVA is used to determine if all the sites have the same mean number of inches of rain. This would be the null hypothesis. The alternative hypothesis is that at least one site has an average rainfall that is different from the others. The company's analysis found the averages at each of the twenty county sites; now they need to see if those averages meet their guess of four inches of rain.

This study meets all the requirements of a good study: there are more than four small sample sets to average, and there is a one-way direction to the data: positive only. (As we said, there can't be negative rainfall.)

In this case, ANOVA is being used to prove or disprove a hypothesis. The null hypothesis is "The average rain on each of our farms is four inches per month or greater." They've found the average of twenty locations they farm in, and they've used ANOVA to find the average of these averages to come up with the rainfall. They can then say, "The average rainfall for our farm sites is four inches a month."

Since one-way analysis of variance is very complex mathematically and is best done with software, you will need to take the time to learn how to analyze the results of the test. That is well beyond the scope of this book.

QUANTITATIVE RESEARCH DESIGN

Setting Up an Experiment That Works

Now you're at the point where you're ready to do a statistical analysis of a problem. Knowing about statistics is good, making them work for you is better, and finding a real-life application for them is best. If you get a chance to work on a quantitative research project, it will look something like what's covered in this section.

Quantitative research design incorporates all of the tools of statistics, but with an emphasis on the results to be used for academic, business, financial, educational, or medical research. We've been talking in this book about the tools that are designed to help you analyze large groups of data in such a study.

The first step is deciding upon the basis for the research: the question that your research will be attempting to answer. When deciding upon a question, keep in mind that it must broad enough to be able to find enough researchable data. Statistics is a data-rich science, and if you've chosen a narrow subject, one that is too specialized, then there won't be enough data. Asking a broad, researchable question is the key to successful quantitative research design.

If you've decided upon a question, then think about a possible answer to that question. Let's look at a medical study:

AN EXAMPLE OF A RESEARCH STUDY

You've decided to measure the comfort and pain-relieving effects of two competitive brands of over-the-counter pain relievers. They have nearly the same price, and they are both sold at one of the largest pharmacy chains in the country. Also, the ingredients in both medicines are the same active ingredients given in clinics and hospitals across the country.

You want to find out which of the two medications gives the most comfort and pain relief to users. You will be testing only those individuals whose pain is at a level for which these two types of medication will work, not a pain level for which a prescription painkiller is required.

Next, you'll need to come up with a possible answer. From your personal and nonscientific observations, your answer is, "Medication A has a 75 percent higher rating for pain relief than medication B." This answer is your null hypothesis.

You now have the question and the answer that you are testing for. The many people across the country who are using the pain medication constitute the population of the study.

SETTING UP THE EXPERIMENT

After making some calls to the marketing department of the drugstore, you discover that 60,000 units of the medications combined are sold yearly. Using computer software you determine that to have a 95 percent confidence level you'll need to create a sample population of 400 people. With these 400, you'll be able to adequately measure what medication works best.

With this sort of study, you've decided to send out online surveys. You build a questionnaire that includes blind questions (that is, questions that help you rule out if the answer is biased because of preknowledge—for instance, if those taking the survey are medical professionals). The questionnaires go out, and the responses come back over the next few weeks.

Levels of Significance

Whether one uses a level of significance of 10 percent, 5 percent, 1 percent, or some other value is dictated by the consequences of being wrong. This is what the level of significance means. You can never be 100 percent correct unless you can get observations from 100 percent of the population. This is just not possible because of financial issues or possibly the fate of using all of the material being tested. (For example, one cannot use 100 percent of the vaccine being tested—if the vaccine works, it's all been used and you have nothing to see. If it doesn't, well, that's too dreadful to think about.)

You next take the responses you've received and determine with statistical analysis that after trying both medications, 82 percent of the responders report that medication A works better on their pain. You've done your data gathering in a random method. You've also determined that the number of responses exceeds the required 400, so you know the sample set is large enough to offer you a 5 percent degree of error that you do not reject your null hypothesis that 75 percent of over-the-counter pain medication buyers have their aches and pains treated more effectively by medication A.

Here are the steps for a quantitative research analysis:

- Ask the question
- Come up with a possible answer: this is your null hypothesis
- Estimate the population size
- Discover the ideal sample size
- Collect the data in a random manner
- Use statistics to match the results to your null hypothesis
- Reject or do not reject your null hypothesis

QUALITY OF THE DATA

Questions and Answers of Data Collection

In this section you will learn the importance of quality data. You will also learn the importance of the elements of quality in the data collection part of your study. These elements include obtaining a randomly generated, repeatable sample set that is large enough to give you the margin of error you require for the study.

In any statistical study the quality of the data is key. Remember, you are looking at a sample of a large population. In most studies, the population is too large for you to observe each one separately and record what you see. In fact, some populations are so large that even the most sophisticated software wouldn't be able to handle it. Therefore, you use samples of the population. Samples can do the trick with statistics—with the right sample, you can use statistics to draw a conclusion as to whether your hypothesis is correct or not. In fact, the whole study of statistics amounts to learning how to use sample data and make an inference about that data.

QUALITY AND RANDOMNESS ARE KEY TO DATA COLLECTION

The most important things about data are the quality and the randomness. Data that is collected in place of an entire population must be very random. It wouldn't do you any good to test only a concentrated part of the entire group. Why? Because this would skew the results of the test. If you conduct a survey using nonrandom information, you run the risk of observing a group that may have a bias. If the group is concentrated in one area or one sector of the entire population, they'll probably have similar answers in your survey.

How do you avoid this? By making the sample set as random as possible. Keep in mind that the sources of the data, the times, and other factors could inadvertently group the sample such that all respondents carry the same bias. The randomness of the sample set gives you the quality of the data. You'll also need to refer to the question and the null hypothesis to help you see if there is a weakness in your data collection. This is best done with questions that are put into the study that would show a bias.

For example, in a survey of a voter's opinion about specific government issues, one of the early questions might be "Do you agree with the president on his agenda?" This could easily bias responders' answers to the question, which align with their feelings about the president in general.

Statistics, while math and software based, still requires a hands-on approach. You'll need to fine-tune the questions, the data collection, the sample size, and the confidence interval. In a perfect world, you would measure and test every single possible combination and every single observable fact in the entire test population. That's the perfect world, and most of the time it can't be done. This is where statistics will allow you to shine: you'll be able to reach conclusions from smaller, easier, more affordable (in terms of time, money, and effort!) tests, and you can use math to back up these conclusions. If done properly and with quality data, you can draw a conclusion that can be duplicated; this is the key to any study.

Surveys

With the increased demand for quality statistics, the creation and assessment of the fairness of surveys and questionnaires is a full-time job.

Quality data can come from many sources. Getting the data into a numbers-based, quantitative format can be easy too, by converting yes/no to an "on/off," "1/0" format, numbering your questions into values from 1–5 (strongly disagree, agree, no opinion, agree, strongly agree), etc. All the data you collect must be put into a format that software can read. If it is just words, the software won't know what to make of it. If, on the other hand, you say, "I liked it" from 1–5, the software will know that someone who answered 5 liked it more than someone who answered 2. The same for yes/no. A no is basically 0 because it didn't happen, while a 1 means it did happen.

QUALITY ISSUES

You can see if your data has quality issues by putting it into a chart. If you use a scatterplot and the scatters have no pattern, then you may be running into a problem with the quality. Why? Because there is no pattern, there is no average, and therefore the data is not measurable. A regression analysis will show this quickly, as the indicators will show that the variable is inconclusive in a relationship. In other words, your regression will show that one variable has no effect on the other. These variables are then not included in the modeling. This is done because after plotting the points on a graph with a scatterplot, you can't draw a line estimate to calculate an accurate slope of the relationship of how one affects the other. This means the data was just too random to have a mathematically based strong relationship that shows how A affects B.

Remember, your data should be randomly collected, but you are looking for one event affecting the other; you are looking for a relationship between the two. The first step, therefore, is collecting the random data so it's not biased. Next you look for how one variable affects the other. This is important, because with this information you'll be able to reject the null hypothesis or not reject it.

This is a great point in a study: to accept or reject the null hypothesis mathematically is quite an achievement. This is particularly the case when setting up a study, designing the survey or data collection, and running the statistical tests has taken weeks or months. If a study such as this ends with a definite "accept" or "reject" that can be proven statistically, the study is a success.

Especially because of this, the quality of the data is important. This is because with any study, you want accuracy to the degree of the margin of error at which you're testing, and this can be achieved only with a randomly generated sample set of a big enough size.

QUANTITY AND SOURCING OF THE DATA

Knowing Where to Look

There are two forms of data collection for any study: qualitative and quantitative. Qualitative data is soft information. This is the information you get from the library or through accessing material on websites. With qualitative studies you are looking at the words. This information is fact, but it's presented in words; *it's not numbers based.*

Qualitative studies are often an integral part of higher educational studies, marketing studies, and sales-related studies. They are often also part of a graduate degree final paper, both at the master's level and at the doctorate level.

The Level of Your Study

The quality of your data is dependent upon the level of your study. If you're running a quick test that serves as a generalization, then it's okay to get data from less reliable sources. Of course, if you are doing a more complex or more critical study, then only use the highest-quality sources of databases.

In the quantitative section of a study, you'll need to have access to numbers-based information. This can come from three sources, in order of quality. The first type of data is collected from other studies that came before you. It is totally feasible to conduct a statistical study using data that has already been gathered by other researchers in the field.

For instance, let's say you're conducting research on whether a certain type of investment product, which is very popular among financial advisers, is in fact a good deal for investors. The same subject has been studied by professional researchers, and the results have been published. The first step is to go to the library and familiarize yourself with all that the experts are saying on the subject, where they got their information, and what statistical tests they ran to come to their conclusions.

You copy the information sources and use the same statistical procedures that the experts used. In effect, you're redoing the experiment. You may not come to different conclusions—in fact you probably won't—but you can reinforce the conclusions others have already drawn. You use the experts' past work as a baseline, a place from which to launch your new study.

The next type of data is that obtained from trusted sources. There are many highly respected databases that are available at a cost; these include www.lexisnexis.com/en-us/gateway.page or www.bloomberg.com/professional/. These sites cost quite a bit, so you'll need to see if your school or library has access to them. Often, you can use your local public library card to open an account at a library in a major metropolitan area and use those databases remotely. Other helpful databases include the following:

- Chicago Public Library: www.chipublib.org/
- New York Public Library: www.nypl.org/
- The British Library: www.bl.uk/
- US Library of Congress: https://loc.gov/
- World Bank open database: https://data.worldbank.org/
- International Monetary Fund database portal: www.imf.org/en/Data

- Bank for International Settlements statistics portal: www.bis.org/statistics/
- World central banks guide website listing: www.centralbanksguide.com/central+banks+list/
- US Census Bureau database: www.census.gov/

The last type of quantitative data is *primary data*. This is data that you've collected from your own observations, from your own direct experiments, or from sending out your own questionnaires.

Professional Databases

As far as databases go, some can be quite expensive, such as www.IBISworld.com. Some of these can be easily replaced with governmental databases. Keep in mind that data collection is a business in itself, and paid-for professional databases, while more expensive, are often easier to use than the governmental databases.

HOW MUCH DO YOU NEED?

You must have enough data to get the numbers right. As we said earlier, if you're working with a sample set smaller than 100, you're running the risk of having a data set that is biased in nature. This is because a sample set that is less than 100 observations has difficulties with randomness.

APPROPRIATE SURVEY DESIGN

Increasing the Efficiency of Questionnaires

As a rule, the questionnaire should be kept short—no more than fifteen or twenty questions. If it gets any longer, then you run the risk of people not finishing, or worse, not responding at all. This is true in face-to-face interviews or phone interviews; people can get irritated by long question sessions. Remember, you are looking for accurate, complete answers to your questions. If your questionnaire is too long and irritates or frustrates those who are taking it, then you will have a lot of unfinished and unusable questionnaires. Worse, if the people start to get frustrated with the length of the questions, they will answer without thinking to finish quickly, or they may answer questions negatively as a reflection of their current mood and not as a true indication of how they feel on the subject.

The only type of questions that should be asked are objective ones. Avoid open-ended questions on the survey and refrain from asking fill-in-the-blank questions. These are bad for two reasons. First, people tend to avoid these types of questions. Second, responses can be very difficult to score or convert into a numbers-based system required for mathematics-based statistics. They also add time to grading and scoring.

You should build into each question a negative-to-positive or an infrequent-to-frequent response with at least seven levels. Seven levels of response will give the test taker enough leeway to answer how he or she would like. Also, seven levels will offer you enough data to run a statistical test of the sample set—and it will provide enough variation to have meaning with a regression analysis, etc.

The item might read, "I've had success with using over-the-counter cold medications" from 1 (least true)–7 (most true).

Paying for Surveys

Should you pay people to take your surveys? Yes and no: you might need to incentivize the people to take the survey, but at the same time it could create bias in the sample. Why? Because the payment may attract a positive bias to the questions. You'll need to think of ways to incentivize survey takers without biasing them.

In these types of studies, it's most common to run a multiple regression analysis after the descriptive statistics tests are done, i.e., the mean and the mode have been determined, and you know the shape of the bell curve. Now the real work begins with inferential statistics and building a predictive model—and the workhorses of these are the regression analyses and the models that can be made after they are done.

Regression is best done with lots of data, and a variation of 7 to 9 data points per question will allow for a good read with it. This will allow enough data to draw a good conclusion of how the independent variables (the questions) affect the dependent elements.

The questions should be set up in clear formats.

It is good to remember that your questions, while looking for soft or qualitative answers, should be built in a way that you can convert the data easily to numbers—a range of answers to a question is best, as this can then be converted to the next step of a statistical study: the numbers-based analysis. For example, "Do you buy your groceries at more than one store? Why?" "Better quality" (1 = less true—7 = more true); "Lower cost" (1 = less true—7 = more true), etc.

THE ETHICS OF STATISTICS

Bending the Truth

Because statistics usually do not stand in a vacuum; because most statistical studies are done in order to prove or disprove something; and because statistics, while math based, rely heavily on interpretation, there is always the chance that people will use statistics to bend the truth.

Data-based sciences that take data, analyze it, and then make predictions based on it are always subject to controversy. We've already mentioned Disraeli's famous dictum about lies, damned lies, and statistics. Numbers can be bent or their interpretation twisted. After all, the numbers may be objective, but their interpretation is in the hands of humans, who are sometimes tempted to get the model to fit the desired outcome.

Do the Right Thing

Ethics questions are easy to answer when you're not personally involved. At the same time, such questions can be very difficult for some people when it involves their work, their reputation, their income, or their status in their community. You hope you'd do the right thing, but keep in mind, people do struggle with this subject when it actually happens to them.

Keep in mind that interpretative sciences are much like an art, and statistical studies and their makers are like art critics. Is art good? Does it have beauty? Sometimes you can even ask the question, "What does it matters if the art is good? That painting sold for one million dollars!" If it's good, pretty, or accurate doesn't matter: the artist somehow convinced the buyer that it was good and worth the one million dollars. Such decisions are often highly subjective (which is why painters often go in and out of fashion). We like or dislike a piece of art often in large part based on prevailing cultural norms. In the same way, we interpret statistics based often on prevailing prejudices.

This is what you must think of when you are doing your statistical studies. Or better yet, this is what you must think of when you are looking at someone else's statistical studies. There is always a temptation to try to get the right numbers, to prove your hypothesis was correct. Sometimes it's unconscious. There is a lot of wiggle room in statistics. The margin of error may be too large to make realistic conclusions, yet the test is called "good" by the testers, and the results are published.

This wiggle room in statistics usually comes in three ways: the collection of the data, the exclusion of outliers, and the overzealous use of wide-parameter margins of error.

Ethics

Ethics are taught more in some subjects than others. Some studies never touch on the subject. Sometimes people who aren't trained in ethics don't even know there is an ethical issue involved. They might not even know they aren't doing the normal, right thing. This often happens when someone is new, or hasn't seen how ethics can affect themselves and others. They might be too young, too inexperienced, or too new to the study.

Collecting the data can cause ethical problems if the sample set is known to be tainted, known to be not random enough, or biased. You may be tempted to oversmooth the data. Smoothing techniques allow you to look at larger groups that show irregularities, but multiple smoothing techniques can destroy the integrity of the data, eliminate the markings of outliers, and make impossible an accurate interpretation of data. A statistical analysis with all the rough edges sanded off and the data smoothed to the point of uniformity may look pleasing, but it's also probably inaccurate.

Finally, the margin of error may be too large. Saying, "The weather next Wednesday will be cold and snowy with a 50 percent margin of error" is meaningless. You're just as likely to be wrong as you are to be right. In this book, we've learned the general guidelines for an acceptable margin of error.

TWO PEOPLE, DIFFERENT INTERPRETATIONS

Two people can look at the same data and come to different conclusions. This happens often with regression analysis, where multiple variables are tested for how they affect another. After a regression is made, the statistician has to interpret what variables affect the other. The software won't tell you this. You have to use your own judgment as to what affects the variable. This can come down to an issue of unconscious bias. You interpret the data in a particular way because you were predisposed to do so—even if you didn't know it.

Looking clearly and objectively at data is a challenge. You are the one to judge the outcomes of the studies you are performing, and you are the one who will be embarrassed if you've done something wrong (or wasted time with an expensive study that went nowhere). On the other hand, you may profit (in fame, financially, or otherwise) from your conclusions. All you can do is be as objective and truthful as possible.

This brings us to the subject of false conclusions. Sometimes, tests are made, and there seems to be a strong direct tie between one thing affecting the other. Without thinking it through, you could draw a conclusion that one affected the other. A good example of this is that in the past twenty years, violent crime in the United States has decreased, while personal computer usage has increased. Someone looking only at those two statistics might draw the conclusion that computer usage affected crime. This is probably not the case (there are all sorts of other factors to be taken into account, including poverty, cultural norms, and so on), but if you weren't careful, you could make this type of false cause/effect assumption.

BIG DATA, SUPERCOMPUTERS, AND ARTIFICIAL INTELLIGENCE

The Future of Statistics and Data Analytics

When statisticians get together over drinks, one of the issues that comes up is whether statistics will move from measuring sample sets to measuring large populations. This is called *big data*. Big data refers to data sets that have hundreds of millions or even billions of data points, rather than the hundreds or even thousands of data points we consider now. Big data, because of its huge size, requires specialized computer memory. After the problem of data storage is taken care of, there comes the problem of running the statistical calculations with a large enough sample set. Remember, with a population in the thousands, an appropriate sample size is in the hundreds. With that in mind, you can imagine an appropriate sample size for a population that is in the billions!

Big Data in the Headlines

Big data and analytics can be a touchy subject. In 2018, Mark Zuckerberg (1984–), the CEO and founder of the social media site *Facebook*, had to go before a US Senate hearing committee to discuss his company's data collection methods. The way in which companies collect and store data about their customers continues to be big news: data collection and the selling of the data to third parties is highly controversial.

Running a statistical model with that much data takes a lot of time and a very powerful computer. Fortunately, researchers and technicians today are building supercomputers. One of the most famous of these companies is Cray (www.cray.com/), which builds computers with the ability to crunch millions of numbers in a fraction of a second. Computing speed and power have big implications for the future of statistics because of this move toward big data. It only makes sense that now that the data is there, companies are willing to invest in human, time, and money capital to use this data for outcomes.

That raises the question: what is all this big data being used for? Today, business is one of the largest users of this information. Government is another. The National Security Agency, the electronic wing of the intelligence community, collects trillions (if not quadrillions) of bits of electronic data, which must be stored and analyzed.

The Popularity of Big Data

What's making data collection so popular? What's making big data possible? It's the advance of speedier computers coupled with cheaper and larger amounts of digital storage. Both are needed for large project data analysis, and in the past few years both have increased in effectiveness while decreasing in investment costs.

What do they use it for? Statistics hasn't really changed: statisticians take sample sets of larger populations and use this information to try to predict the future. That's about all they are really doing: collecting data, crunching numbers with supercomputers, applying statistics, and attempting to predict the future. Some try to predict consumer trends, some try to predict the direction of the stock market, some try to predict the outcomes of elections.

Knowing what your customer will ask for at the market and being there first with it can give you the competitive edge, and with this information you can (in theory) make more money. Knowing what issues resound with the electorate means you can fine-tune your candidate's speeches to hit those points and increase your chances of winning an election.

Big data is the wave of the future. In turn, this leads to the development of new computer languages to statistically crunch numbers in multiple sequence. These languages allow the numbers to crunch at an even faster pace than sequential crunching.

Another influence on statistics today is in the field of chaos theory. Chaos theory is a separate division of mathematics and physics that deals with extreme randomness. It has significant applications in the area of quantum mechanics, the far reaches of theoretical physics.

In the field of statistics, ever-growing computer power is the future. Where this will eventually lead is anyone's guess.

INDEX

C

D

O

P

Q

R

S

T

V

W

Y

Z

ABOUT THE AUTHOR

David Borman is a professor of economics, accounting, and data at a Chicago university. He has worked in finance and accounting since 1999, including working at Morgan Stanley, Phillip Capital, and Custom House Global Fund Services. He has a BS in finance and an MS in accounting, and is working on his PhD in financial management. Borman has studied financial and statistical modeling at Northwestern University, Wharton, and MIT. He is the author of *Day Trading 101, The Everything® Guide to Currency Trading, The Everything® Guide to Commodity Trading,* and more.